WATER
CONSERVATION, REUSE, AND RECYCLING

PROCEEDINGS OF AN IRANIAN-AMERICAN WORKSHOP

Committee on US–Iranian Workshop on
Water Conservation and Recycling

In cooperation with the Academy of Sciences of the Islamic Republic of Iran

Office for Central Europe and Eurasia
Development, Security, and Cooperation
Policy and Global Affairs
NATIONAL RESEARCH COUNCIL
OF THE NATIONAL ACADEMIES

THE NATIONAL ACADEMIES PRESS
Washington, D.C.
www.nap.edu

THE NATIONAL ACADEMIES PRESS 500 Fifth Street, N.W. Washington, DC 20001

NOTICE: The project that is the subject of this report was approved by the Governing Board of the National Research Council, whose members are drawn from the councils of the National Academy of Sciences, the National Academy of Engineering, and the Institute of Medicine. The members of the committee responsible for the report were chosen for their special competences and with regard for appropriate balance.

This study was supported by the National Research Council and the W. Alton Jones Foundation. Any opinions, findings, conclusions, or recommendations expressed in this publication are those of the author(s) and do not necessarily reflect the views of the organizations or agencies that provided support for the project.

International Standard Book Number 0-309-09293-0 (Book)
International Standard Book Number 0-309-54502-1 (PDF)

One copy of this report per request is available from the Office for Central Europe and Eurasia, National Research Council, 500 Fifth Street, N.W., Washington, DC 20001; (202) 334 2644; Fax, (202) 334 2614.

Additional copies of this report are available from the National Academies Press, 500 Fifth Street, N.W., Lockbox 285, Washington, DC 20055; (800) 624-6242 or (202) 334-3313 (in the Washington metropolitan area); Internet, http://www.nap.edu

Printed in the United States of America

THE NATIONAL ACADEMIES

Advisers to the Nation on Science, Engineering, and Medicine

The **National Academy of Sciences** is a private, nonprofit, self-perpetuating society of distinguished scholars engaged in scientific and engineering research, dedicated to the furtherance of science and technology and to their use for the general welfare. Upon the authority of the charter granted to it by the Congress in 1863, the Academy has a mandate that requires it to advise the federal government on scientific and technical matters. Dr. Bruce M. Alberts is president of the National Academy of Sciences.

The **National Academy of Engineering** was established in 1964, under the charter of the National Academy of Sciences, as a parallel organization of outstanding engineers. It is autonomous in its administration and in the selection of its members, sharing with the National Academy of Sciences the responsibility for advising the federal government. The National Academy of Engineering also sponsors engineering programs aimed at meeting national needs, encourages education and research, and recognizes the superior achievements of engineers. Dr. Wm. A. Wulf is president of the National Academy of Engineering.

The **Institute of Medicine** was established in 1970 by the National Academy of Sciences to secure the services of eminent members of appropriate professions in the examination of policy matters pertaining to the health of the public. The Institute acts under the responsibility given to the National Academy of Sciences by its congressional charter to be an adviser to the federal government and, upon its own initiative, to identify issues of medical care, research, and education. Dr. Harvey V. Fineberg is president of the Institute of Medicine.

The **National Research Council** was organized by the National Academy of Sciences in 1916 to associate the broad community of science and technology with the Academy's purposes of furthering knowledge and advising the federal government. Functioning in accordance with general policies determined by the Academy, the Council has become the principal operating agency of both the National Academy of Sciences and the National Academy of Engineering in providing services to the government, the public, and the scientific and engineering communities. The Council is administered jointly by both Academies and the Institute of Medicine. Dr. Bruce M. Alberts and Dr. Wm. A. Wulf are chair and vice chair, respectively, of the National Research Council.

www.national-academies.org

NRC COMMITTEE ON US–IRANIAN WORKSHOP ON WATER CONSERVATION AND RECYCLING*

Henry Vaux, *Chair*
Associate Vice President and Professor
Division of Agriculture and Natural Resources
University of California

James Crook
Principal Water Reuse Technologist
Global Water Business group
CH2M Hill

Rebecca Parkin
Associate Research Professor
Department of Occupational and Environmental Health
George Washington University

Staff

Glenn Schweitzer
Project Director
National Research Council

Sara Gray
Senior Program Associate
National Research Council

Christopher Holt
Senior Program Assistant
National Research Council

Amy Moore
Program Assistant
National Research Council

---

*Titles as of December 2003.

Preface

In December 2002, a group of specialists on water resources from the United States and Iran met in Tunis, Tunisia, for an interacademy workshop on water resources management, conservation, and recycling. This was the fourth interacademy workshop on a variety of topics held in 2002, the first year of such workshops. Tunis was selected as the location for the workshop in order to simplify travel arrangements after political issues complicated the holding of the workshop in California. Also, the Tunisian experience in addressing water conservation issues was of interest to the participants from both the United States and Iran.

The National Academy of Sciences and the National Academy of Engineering selected the American participants. The Iranian Academy of Sciences of the Islamic Republic of Iran selected the Iranian participants. Several Tunisian specialists also attended the workshop at the invitation of the National Academies. All attendees participated in their personal capacities, and the papers that were presented and the related comments represented their personal views.

This report includes the agenda for the workshop, all of the papers that were presented, and the list of site visits. In order to encourage open and candid discussion at the workshop, no record was kept of the comments during the discussion periods.

The American participants were from the arid and semi-arid portions of the western United States. Iran is predominantly an arid and semi-arid land. Thus the symposium participants had a common context of experience with water resources and interests in water management and sustainability. Indeed, surprising similarities between the water situation in Iran and that in the United States quickly became evident. The following points of similarity emerged:

- Population pressures on water resources grew rapidly in the last half of the 20th century;
- Rapidly urbanizing populations are changing the essentially agrarian nature of many areas;
- Economic growth is placing new demands on available water supplies;
- Agriculture is by far the largest user of available water;
- The geographical distribution of water does not match the geographical distribution of population;
- Use of ground water is increasing rapidly with an increasing ground water overdraft;
- Storage and conveyance facilities have long been used to manage water scarcity;
- The need for new management regimes in an era of intensifying scarcity is recognized;
- The importance of stakeholder participation in fashioning solutions to water problems is acknowledged.

In addition to problems of water allocation, each country is confronted with problems of climatological extremes. Efforts to manage droughts and floods are similar in the two countries and have had similar results.

All of these similarities mirror in a general way the water problems found in arid and semi-arid regions throughout the world.

Although Iran and the United States have somewhat different histories in their efforts to cope with water scarcity, the efforts of both are instructive. Each country is at a different stage along a trajectory of efforts to solve water problems. Each country has had successes and each has had failures. The papers in this volume address experiences of each country and identify water management problems that must receive high priority in the future.

At the conclusion of the conference, participants discussed the most mutually pressing water problems. There was agreement on the four most critical problems:

- Forecasting and managing droughts.
- Developing technology for inexpensive recycling of urban wastewater without adverse impacts on public health.
- Improving the economic efficiency in using water in agriculture.
- Developing new and innovative institutional arrangements for managing water consistent with historical antecedents and traditions of each country.

Following the workshop, there has been correspondence between the participants concerning future opportunities for consultations and workshops.

Acknowledgments

The specialists and officials that participated in this activity are identified in the appendixes. Their contributions are greatly appreciated. Mehdi N. Bahadori, Vice-President for Research of the Iranian Academy of Sciences, deserves particular recognition for his continuous efforts in ensuring that the activities would be professionally rewarding for water resource specialists.

We wish to extend special thanks to Ameur Horchani, Tunisian Secretary of Agriculture, and to Lotfi Ghedira, Director of the Bureau des Équilibres et de la Planification Hydrique of the Tunisian Ministry of Agriculture, for their special attention to the arrangement of site visits and participant visas.

Special appreciation is extended to the National Research Council and the W. Alton Jones Foundation which provided funding for this project.

This volume has been reviewed in draft form by individuals chosen for their technical expertise, in accordance with procedures approved by the NRC's Report Review Committee. The purpose of this independent review is to provide candid and critical comments that will assist the institution in making its published report as sound as possible and to ensure that the report meets institutional standards for quality. The review comments and draft manuscript remain confidential to protect the integrity of the process.

We wish to thank the following individuals for their review of selected papers: Richard Foltz, University of Florida; Charles Howard, Charles Howard and Associates; John Letey, University of California at Riverside; Miguel Marino, University of California at Davis; and Michael Stenstrom, University of California at Los Angeles.

Although the reviewers listed above have provided constructive comments and suggestions, they were not asked to endorse the content of the individual papers. Responsibility for the final content of the papers rests with the individual authors.

We wish to thank Aws Alouini, Lowell Lewis, Sara Gray, Christopher Holt, and Amy Moore for their assistance during the project.

—Henry Vaux
Chair, NRC Committee on US–Iranian Workshop on Water Conservation and Recycling

—Glenn E. Schweitzer
Director, Office for Central Europe and Eurasia, NRC

Contents

WATER MANAGEMENT IN THE UNITED STATES AND IRAN

Water Management, Conservation, and Reuse in the Western United States

Henry Vaux, Jr.

INTRODUCTION

In the United States the patterns of precipitation, water use, and water management problems are highly variable. In examining these patterns it is common to divide the country into relatively dry and wet regions at the 100th meridian. By coincidence, the 100th meridian is the line that separates the relatively moist eastern region of the country where rainfall exceeds 500 mm annually from the relatively arid western portion of the country where, with some exceptions, annual rainfall is less than 500 mm annually. As a general rule, agriculture cannot be carried out productively without supplemental irrigation unless precipitation exceeds 500 mm annually. The hydrologic circumstances and the problems of water management tend to be somewhat different in the relatively moist eastern portion of the country than they are in the arid and semiarid regions. In the moister regions, the focus tends to be on maintaining and enhancing the quality of water. Periodic droughts cause times of water scarcity, but as a general rule scarcity tends not to be as significant a problem in the eastern United States as it is in the West.

By contrast, in the semiarid and arid western portions of the country, water scarcity tends to be the predominant water problem and water quality has been managed more carefully for longer periods of time simply as one means of preventing scarcity from intensifying. Although the western United States and particularly the southern portion of the West is often characterized as the place of wide open spaces, for the last decade or so it has been the most urbanized portion of the country. The urbanization of the southwest continues as many of the cities there (for instance, Albuquerque, Denver, Las Vegas, Los Angeles,

Phoenix, and Salt Lake City) are among the fastest growing cities in the United States.

This paper is intended to provide an overview of the water situation in the United States. However, the focus will be on the semiarid western portions of the country since the hydrologic circumstances and problems there tend to be more closely akin to those encountered in Iran. The broad overview that follows is divided into five sections. The first of these explores the problems of water scarcity and describes some of the mechanisms that have been developed to manage scarcity. The second section focuses on water quality and describes the strategies that have been used to address water quality issues in the United States. The third section focuses on groundwater and groundwater management. Groundwater accounts for a significant portion of the water supply in most regions of the West. While the lessons of groundwater management are reasonably well known, efforts to apply those lessons have not always been successful. The final two sections focus on the problems of consumptive and nonconsumptive uses, respectively. Early efforts to develop and manage the water resources of the West were focused on acquiring sufficient water to serve consumptive uses. In the last two or three decades instream uses or nonconsumptive uses, particularly those related to the maintenance and enhancement of the environment, have become quite important, and this has led to efforts to rebalance the quantities of water allocated among consumptive and nonconsumptive uses.

Water recycling and reuse is one of the main focuses of interest here. Recycling and reuse is becoming increasingly important within each of the subject matter domains to be discussed in this paper. As a consequence, an effort has been made to emphasize the current and anticipated future role of reclaimed and reused water in managing scarcity, preserving and enhancing water quality, managing groundwater, and in the management of consumptive and nonconsumptive uses.

WATER QUANTITY: THE PROBLEM OF SCARCITY

Throughout the western United States water supplies are scarce. Simply stated, this means that there is not as much water as people wished there were. There is not enough water to fulfill simultaneously all of the wants for water for agricultural, domestic, industrial, and environmental uses. The bases of this scarcity are several and are more complicated than just a simple lack of water. The occurrence of water over the western landscape is variable both in time and location. Virtually all of the West has climates that include both wet seasons and dry seasons. Thus, for example, California has a Mediterranean climate with a dry summer and wet winter season. Much of Arizona and neighboring areas have two wet seasons, a rainy winter period and a short summer monsoon period. Here, precipitation, which is not great to begin with, can be concentrated in five or six rainfall events. Similarly, throughout much of the West, places where water occurs plentifully often are not the same as places where water is used. Much of the water supply is generated in the form of snow pack in the higher elevations while the places of use tend to be in the coastal zones and broad lowland valleys.

Historically, these imbalances between the times and places where water occurs and the times and places where it is used were redressed by developing facilities that would allow water to be captured and stored in wet times and places and transported to places of use in dry times. The result is that one of the prominent characteristics of today's western waterscape is a vast maze of dams and canals. Parts of this maze are so imposing physically that they can be seen from outer space. The storage and conveyance facilities of the western United States have converted a portion of the highly variable supply into firmer supplies some of which can be "guaranteed in all but the most severe of drought conditions." This, in turn, has allowed major cities such as Los Angeles and Phoenix to grow far beyond the limits imposed by local water supplies and have resulted in a vast and extensive irrigated agriculture that supplies significant portions of the nation's food and fiber demands at all seasons. Available water supplies have also allowed the region to develop industrially and have supported historically such important industries as primary metal production and aircraft fabrication. Today, the fabrication of computer chips and the manufacture of equipment that is dependent upon semiconductors requires a reliable supply of high quality water.

Robust economic and population growth has continued almost without a pause in the western United States for more than 100 years. During all of this time water scarcity has continued to intensify as the demands for water to support larger populations, more and more extensive irrigated agriculture, more sophisticated and higher valued industry, and recreational purposes have grown far more than in proportion to the population. Scarcity is exacerbated by outmoded water laws and institutions, many of which were designed to address conditions that prevailed in the 19th and early 20th centuries. The need to modernize institutions has been underscored in recent decades as it has become clear that it is not possible to address water scarcity exclusively or even primarily by continuing to build impoundment and transport facilities. There are several reasons for this. First, the attractive damsites that are relatively close to centers of agriculture and population have already been developed, and what remains are damsites that are either difficult and costly to develop or so remote from places of use as to make water conveyance financially infeasible. Second, civil works such as dams and canals have become more expensive because of engineering difficulties and remoteness even while the competition for public funds to pay for such facilities has become far more acute than in the past. Increasingly, in the last five decades, government in the United States has been called upon to provide a vast array of new social services and regulatory support. The result is that public funds are not as easily available to support civil works as they were in the past. Third, it is now recognized that the environmental consequences of dams and canals are more serious and far reaching than was thought to be the case during the heyday of dam building. By altering flows, water temperatures, and other key environmental parameters, dams and diversions are now understood to contribute to the loss of biodiversity, to impair fish habitat, and to constrain the capacity of the environment to provide waste assimilation and other services.

The consequence is that water managers in the West have turned to a broad array of techniques to assist in the management of scarcity. Many of these techniques are subsumed under the title of conservation and demand management. Their purpose is to regulate and limit the use of water thereby bringing demand more into balance with supplies. One way in which this is done is by adopting and employing technology that makes more efficient use of water. Flow constrictors and low flush toilets can reduce domestic uses substantially. Closed conduit irrigation systems that allow for precise application of water and reduce losses due to soil nonuniformity have resulted in substantial savings in the water needed to irrigate certain crops. Similarly, water efficient industrial technology permits the manufacture of all manner of commodities at significant water savings.

Programs of education have also been successful in reducing the demand for water, particularly in the domestic and agricultural sectors. Domestic users tend to be more careful and use less water when they know how much water they actually use. These users also respond to education programs that explain where water comes from and elaborates on the problems of managing it. Similarly, agricultural water users tend to practice more careful water management when they know with some accuracy how much water they use. Programs to educate agricultural users in the benefits and techniques of appropriate irrigation scheduling and various means of managing water on nonuniform soils have also been effective.

Finally, there are some rather straightforward institutional measures that can be employed to manage scarcity. These include pricing and markets. Appropriate pricing of water is also an effective way of managing demand and inducing conservation. Historically, water in the United States has been priced only to recover the costs of treating and transporting it. The water itself has been free and has sometimes been treated as if it were free. Policies that call for the price to more nearly reflect the scarcity value of water have been quite effective in reducing water use, particularly in times of drought. Pricing practices such as progressive rate structures that mimic traditional marginal cost pricing rules have led to widespread conservation both by encouraging adoption of water saving technology and by encouraging more careful and efficient use and management of water. This is true both in households and on the farm. Pricing policies tend to be quite controversial, but there is little doubt that they are effective.

Water markets have also been used successfully to ration large water wants among relatively fixed sources of supply. By relying on voluntary bargaining, markets ensure that both buyer and seller gain in the exchange of water and also help to ensure that water is allocated to its highest valued uses. Markets have been slow to develop because of concern about third party effects and worries that instream uses and common pool uses cannot compete on a comparable basis with consumptive uses. The controversy also extends to concerns that most water to be transferred will come from agriculture and rural areas and that the financial returns from selling or leasing the water may enrich places other than areas of origin. In spite of these concerns, water marketing has proved an excellent means of coping with limited water supplies in specific drought situations and in some regions of the West where markets have been well-established for decades. It seems clear that the full potential of water markets remains to be realized.

Although augmentation of developed supplies still remains an option to meet growing demands for consumptive uses, issues related to financing and environmental impact will likely constrain the extent to which new supplies will be developed in the future. In many regions of the West, the only possible source of additional water supply is reclaimed and recycled water. The Orange County Water District in southern California has been a pioneer in developing recycled water supplies to repel salt water intrusion to local aquifers and to replenish those aquifers that are a source of domestic supply. The city of Phoenix, Arizona, has developed a wastewater recycling plan that will virtually eliminate the discharge of wastewater, treated or not, into an adjacent river. As will be noted later, strict water pollution control laws that embody stringent regulations on what may be discharged to public waterways have provided strong incentives for industrial and municipal wastewater recycling.

Intensifying water scarcity will remain a dominating feature of the western water scene for the foreseeable future. Only the development of low cost sea water desalting technology could provide any substantial relief. Western water managers have had some success in managing water scarcity by utilizing a wide variety of management techniques, including supply augmentation, pricing, education, water saving technology, markets, and water recycling. In the future these techniques together with new technology and innovative management systems will all have to be employed to stretching limited water supplies to serve the growing demands for additional water from almost every water using sector.

MANAGING WATER QUALITY

There is a tendency in the United States to forget that deteriorating water quality can reduce available water supplies just as surely as drought. This is due, in part, to the practice of separating considerations of water quantity from considerations of water quality and the tendency to treat these independently of each other. The fact is that the amount of water available for any purpose in any location is a function of the quality of available water supplies. Thus, it is important to recognize that considerations of water quantity and water quality are intimately related to each other and should be considered jointly with each other. In this paper, they are separated only to simplify and highlight the various issues related to each.

Historically, in the United States, the maintenance and enhancement of water quality was considered to be the prerogative of the states. Western states tended to have reasonably effective water quality control laws and policies because the prevailing water scarcity made clear the importance of not allowing water quality to deteriorate. Despite this fact, there was a tendency nationwide for states to compete for the location of new business by promising freedom from stringent water quality regulations. This provided states with an incentive to weaken water quality policies and regulations so as not to be at a competitive disadvantage when promoting economic growth.

Beginning about 1970, the U.S. Congress passed a series of water pollution control and water quality laws that created strong national programs to restore and maintain water quality. These programs were aimed at reducing (and ultimately eliminating) most point source pollutants. (Point source pollutants are those that are discharged to the environment from a discrete point, for example, an outfall). The federal pollution control effort had two distinct designs, one aimed at controlling industrial discharges to the nation's waterways and the other focused on the management of waste in public sanitary systems.

For point source industrial discharges, a permit system was developed and enforced that requires all firms that discharge to waterways to acquire a permit to do so. In addition, the law required the use of the best available technology (BAT) to treat wastewater prior to discharge. Over the course of the years BAT came to be defined in a way that included an economic test. This test was intended to avoid bankrupting large numbers of firms through regulation. Firms that were not eligible for permits because of failure to employ the best available technology were barred from discharging. The permit discharge system was initially backed by an enforcement system that severely limited its effectiveness. Subsequently, the application of a provision in some older legislation that gave citizens the right to identify and file enforcement actions led to workable regimes of enforcement and resulted in an effective wastewater management strategy. One result of this strategy has been to encourage firms to recycle and reuse both process and cooling water. In the arid and semiarid southwestern United States manufacturing firms typically recycle extensively with the result that the demand for water for industrial purposes has grown at rates far lower than the rates of growth in demand for other purposes.

The companion strategy and program that was applied to the control of wastewater from public sanitary systems was an immense civil works program in which the federal government subsidized 75 percent of the costs of constructing wastewater treatment systems. Initially, such systems were required by law to include at least secondary treatment. Today, most such systems include some form of Advanced Wastewater Treatment (AWT), which in the past was sometimes called tertiary treatment. The total costs to the federal government of the wastewater treatment subsidization over the nearly 30 years of its existence totals approximately $100 billion. The primary criticism of this strategy is that the large subsidy led to treatment technologies that were unnecessarily capital intensive and delayed the development and adoption of innovative technologies, such as artificial wetlands, beyond the time when they could have first been employed effectively as part of a comprehensive wastewater management strategy. Despite the shortcomings of the permit and technology strategy and the capital intensive public wastewater treatment strategy, the twin strategies have been very effective—even if unnecessarily costly—in managing point source contaminants.

In spite of the success of laws and policies governing the management of discharges to surface waterways, water quality in the United States continues to be subject to deterioration. One explanation lies with the fact that early policies failed to address in any adequate way the problems of pollution from nonpoint sources. For example, agriculture was exempted from early pollution control policies. Moreover, policies aimed at point source pollutants rely for their effectiveness on the fact that points of discharge can be identified and subjected to controls. This is not the case with nonpoint source discharges. Uncontrolled nonpoint source discharges are now the most significant cause of water contamination in the United States.

Nonpoint source discharges affect the quality of both ground and surface water. Groundwater pollution can be especially serious for several reasons. First, as a general rule the self cleansing and diluting properties of groundwater are not nearly as robust or effective as those of surface water. Second, groundwater is particularly susceptible to contamination from discharges made many decades ago. Fertilizer residues, toxic chemicals, and other materials discharged onto and into the soil can pose a serious hazard to groundwater quality for many years. Often such threats cannot be identified until the groundwater is already contaminated. Even in instances where spills and other potential sources of contamination that are already in the soil profile can be identified, the costs of clean-up may be astronomical. In the last 10 to 15 years effective techniques for cleaning up groundwater contamination in situ have been developed, and these hold considerable promise for keeping the costs of clean-up at reasonably manageable levels. Yet, protection of groundwater quality remains one of the major water resource management challenges for the future in the United States.

The struggles of the last several decades to fashion techniques and policies for managing nonpoint source contaminants in the United States now appear to have resulted in the targeting of best management practices (BMPs) and comprehensive watershed management as the two preferred strategies. Best management practices are aimed at specific land use activities while watershed management entails the holistic management of watershed lands in ways that restrict activities to areas that are least likely to contribute to water pollution. Thus, best management practices require the adoption and implementation of a prescribed set of practices for conducting activities such as lumbering and agriculture. The prescribed practices are intended to reduce or eliminate altogether the entrainment of contaminants in surface water runoff or deep percolation through the soil profile. One example of a BMP is the use of closed conduit irrigation technologies that allow precision application of water to avoid overirrigation that contributes to run-off and deep percolation.

Comprehensive watershed management strategies are plans and practices for managing watersheds as a hydrologic unit and regulating activities. The resulting regulations aim to direct activities that can generate nonpoint source contamination to the least susceptible areas of the watershed and provide strong protections for susceptible areas such as riparian zones. Riparian zones offer defenses from contamination originating elsewhere but can themselves be the source of nonpoint source pollution if not carefully managed.

The primary challenge with the implementation of these strategies in the United States is finding an appropriate balance between the personal freedom to choose how to undertake activities such as lumbering and agriculture, and regulations that are effective in controlling nonpoint source pollution. Experience with these strategies is limited, and it is not yet clear whether they will work. Much more experience with them will be required. In the meantime the search for new and innovative ways of managing nonpoint source contaminants will continue.

The semiarid western United States and particularly the more arid southwestern region is subject to soil salinization, which results from irrigation practices. Without proper management, soils can become salinated either from salts introduced with irrigation water or salts mobilized in the soil profile by the presence of irrigation water. In the absence of management, the salinization of soils will lead to a reduction in agricultural productivity and ultimately to sterilization of the soil itself. Proper management requires the application of sufficient quantities of water to leach salts below the root zone and appropriate drainage facilities to carry away leaching waters and prevent the build-up of groundwater tables and the water-logging of soils.

There are only a few places in the western United States where salinity is managed on a sustainable basis. Leaching widely occurs where salt is a problem, and it is sustainable in circumstances where drainage waters can be disposed of or recycled. In many areas, the disposal of drain water constitutes a major problem, and irrigators are unable to find workable options for the management and disposal of drainage waters. In these areas, salinity will remain a persistent problem and unless ways are found to deal with drain water it is likely that irrigated agriculture will ultimately become untenable.

Although the United States has had some successes in managing water quality, the successes are largely found in the management of point source discharges to surface waters. The strategy of regulating industrial discharges with permits and requirements for the use of best available technology, together with the strategy of subsidizing the construction of waste treatment systems to clean sanitary wastewaters, has worked reasonably well. These strategies emphasized the use of regulations and capital intensive treatment regimes, with the result that the use of market-like incentives and innovative treatment techniques were under-utilized. This meant that the success in treating point source discharges was probably more costly than it needed to be.

The effective management of nonpoint source contaminants remains the major water quality challenge for the future. A number of techniques, including the implementation of best management practices and comprehensive watershed management, are available but have been implemented in a limited number of situations. More experience is needed with these techniques. In addition, it seems likely that the nonpoint source pollution problem will be difficult to solve without major new innovative techniques and methods for managing diffuse pollution.

THE MANAGEMENT OF GROUND WATER RESOURCES

Groundwater is a significant source of total water supply in the United States and is the source of approximately 25 percent of the drinking water. Major cities such as San Antonio, Texas, and Tucson, Arizona, are completely dependent on groundwater for drinking water supplies. Throughout the United States, but especially in the West, water managers have been slow to learn how to manage groundwater on a sustainable basis. Aquifers have frequently been excessively overdrafted with the result that pumping depths have become deeper than optimal, pumping costs have risen above affordable levels for some users, and water scarcity has intensified. Historically, the principal strategy for addressing this problem entailed the development of additional surface supplies to offset the decline in economically affordable groundwater. The viability of this strategy has been significantly reduced because of the financial costs of building the dams and canals necessary to provide additional surface supplies and because of adverse environmental impacts associated with these facilities.

Apart from the problems of managing groundwater quality that were discussed in the previous section, the primary problem confronting groundwater managers is that of regulating extractions to ensure economic sustainability. Overdraft is said to exist when the quantities extracted exceed the quantities recharged. Overdraft may be economically efficient in certain situations and thus it is not always bad. However, prolonged or permanent overdraft is always self-terminating as groundwater depths are drawn down to levels from which it is no longer economical to pump. As a general rule, in the absence of regulation groundwater will always be extracted inefficiently when extraction is organized in an individually competitive fashion. In these circumstances, extractors have an incentive to ignore the increments added to future pumping costs by extracting now rather than later. Monopolistic pumpers, to the contrary, have every incentive to account for these costs and thus tend to extract groundwater in an economically efficient fashion.

The problem of efficient groundwater management, then, can be summarized as the problem of managing groups of independent extractors to account for future costs of extraction (the marginal user cost) and behave in the same fashion as a monopolistic extractor would behave. This cannot be accomplished without metering and regulating the extractions of each individual. Quotas and pump taxes are the usual tools recommended for regulation, and quotas are the most often used technique.

Another technique that can be used with or without regulation is to supplement natural rates of recharge through artificial recharge. Artificial recharge can be accomplished either indirectly by percolating water through the soil profile underlying percolation basins or by injecting water directly into the aquifer. By increasing the total quantities of water recharged through artificial recharge, managers can increase the quantities of water that can be extracted sustainably. The possibilities for artificial groundwater recharge also open the prospect of groundwater storage or groundwater banking.

The concept of groundwater banking is very similar to the concept of surface water storage. Surplus flows in wet times can be stored in a groundwater basin and then extracted for use in dry times. Groundwater storage or groundwater banking has several advantages over its surface water counterpart. It avoids the adverse environmental effects of surface water storage systems and the recharge facilities and wells needed to make groundwater storage systems work are generally far less expensive than surface water storage facilities. The possibilities of effecting groundwater storage or banking with reclaimed wastewater are also beginning to be exploited. Artificial groundwater recharge and storage are being employed on a large scale in the water management schemes of Phoenix, Arizona, and Orange County, California, as well as about two dozen other sites around the country.

The lack of effective controls on groundwater extractions can constrain the use of groundwater storage and banking. Thus, it also constrains, often severely, the extent to which ground and surface waters can be managed conjunctively. In the absence of workable extraction controls, the party who pays for the recharge operation and recharges the aquifer has no guarantee that he or she can capture all of the benefits from recharge. Other extractors, who did not participate in the recharge operation, may be able to extract some of the recharge water at no cost other than the cost of extraction. Thus, the lack of effective pumping controls tends to create a barrier to the successful implementation of artificial groundwater recharge and groundwater banking programs. The full promise of groundwater recharge and banking in California as well as the full promise offered by the conjunctive use of ground and surface waters are unlikely to be realized until such time as state or local agencies are able to adopt some effective form of regulating groundwater extractions.

Finally, it is important to point out that the reluctance to implement groundwater management practices that include effective monitoring and control of extractions may be due to the fact that the benefits of such regulation are reasonably small. There are several studies in California that show that the benefits of groundwater regulation in most agricultural areas are relatively modest. By contrast, in two areas in which stringent groundwater management regimes have been developed in the last two decades, the benefits were apparently quite large. In the state of Arizona, the development of a comprehensive groundwater management plan became a quid pro quo for a large federal investment in a surface water storage and conveyance facility called the Central Arizona Project. The thinking here was that in the absence of groundwater controls, the need for additional supplies of surface water would continue almost indefinitely as the state's groundwaters were drawn down to uneconomical pumping depths.

In a similar vein, nearly ten years ago the city of San Antonio, Texas, found its groundwater supplies endangered as a consequence of unrestricted extractions from neighboring users. San Antonio, which is one of the two or three largest cities in the United States that relies exclusively on groundwater, was able to motivate regulatory authorities in the state of Texas to take necessary actions to control extractions, thereby preserving the city's water supply. Here again, the benefits from management and regulation of a groundwater resource were quite high. These experiences suggest that in the United States the existence of positive benefits from groundwater regulation is a necessary but not sufficient condition to induce programs of regulation. Experience shows that the benefits must be very large and quite visible before it is possible to move forward to comprehensive management programs.

The management of groundwater quality and quantity remains a major challenge in the United States. The techniques of managing groundwater quantities are better known and generally less expensive to apply than the techniques for managing groundwater quality. For the most part the management of groundwater remains within the prerogative of the states. One of the significant debates confronting the groundwater management community centers on the question of whether the federal government will have to assume substantial authority if the nation's groundwaters are to be adequately protected.

CONSUMPTIVE USES OF WATER

For the first two hundred years of development in the United States the focus was on the development of sufficient water supplies to serve traditional consumptive uses. Consumptive uses are those that transform the water—either qualitatively or in phase terms—in ways that make it unfit for further utilization. Typical consumptive uses include irrigation, domestic household use, and industrial processes water use. (Industrial thermal uses are partly consumptive but most thermal waters are recooled and made available for other uses.) The narrow historic focus on consumptive uses of water tended to ignore and obscure important instream uses, particularly environmental uses, that were increasingly sacrificed to support the growth in consumptive uses. Today, it is widely recognized that water development and management activities must give balanced consideration to both consumptive and instream uses.

There is a need for new technologies that will allow water to be used more efficiently in all consuming sectors. Yet, there is a surprising lack of knowledge about the fundamental determinants of consumptive use, knowledge that would be needed to develop appropriate technology. Little or nothing is known about the determinants of public and commercial uses. Although these uses are small when compared with industrial, domestic, and agricultural uses, an understanding of the determinants of these uses will be needed if a comprehensive scheme to manage all uses is to be developed in the coming decades. It is well understood that industrial demands for additional water are highly sensitive to pollution control regulations. Nevertheless, the precise nature of the relationship between the quantities of industrial water demanded and pollution control laws has never been documented and specified.

The determinants of domestic use have been identified for a number of large metropolitan areas and are frequently used in devising sophisticated schemes for managing domestic water use. The determinants of domestic water use have not been investigated or specified for most medium and small sized communities. Additionally, it has been more than 30 years since the last comprehensive study of the determinants of domestic use. As water scarcity continues to intensify it will be important to have a comprehensive understanding of the variables that determine domestic consumptive uses so that sophisticated schemes of domestic water supply management can be applied to all communities and not just those who are able to afford the elaborate studies needed to identify the determinants of use.

Generally, water use for irrigation purposes is understood to be dependent upon climatic variables such as temperature and humidity, crop type, and the uniformity with which water is applied. The effect of these variables on levels of agricultural water use has been characterized for some areas but not for others. A complete knowledge of the determinants of agricultural water use is virtually a precondition for the development of efficient and effective schemes of agricultural water management. A comprehensive assessment of determinants of agricultural water use for all regions where irrigated agriculture is practiced will be essential in the future as the supplies of water available to grow food come under intensifying competitive pressure.

Agricultural water uses are especially important both because they account for more than 80 percent of the consumptive use in the United States (more in other parts of the world) and because the need for water to grow food for a growing global population is likely to increase in the coming decades. There will be compelling reasons for learning how to manage water efficiently, to minimize the impacts of fertilizer and nutrient residues on receiving water, and to manage salt balances in precise and realistic ways. The knowledge generated could result in agricultural management practices that make efficient use of scarce water supplies while having minimal impacts on water quality.

There are areas in the United States and elsewhere in the world where irrigated agriculture cannot be extended indefinitely because they depend upon nonreplenishable groundwater supplies or because lands are not suited to agricultural production over the long run. As irrigated lands go out of production it may be possible to carry on profitable dry-land farming. More knowledge is needed about the circumstances under which dry-land farming can be carried out profitably. One possibility is the development of crop varieties that are specially adapted to dry-land conditions and can produce higher yields than might otherwise be expected.

The potential for genetic alternation to improve crop water use deserves further and comprehensive exploration. Crops with different photosynthetic pathways have clear water use differences, and it may be possible to take advantage of such differences in the future. However, it is unclear whether genetic manipulation can be used to achieve changes in the fundamental properties of crop water use. It may turn out to be more productive to focus on genetic manipulation to increase crop rooting depth, improve crop quality, and develop crop characteristics that reduce the need for fertilizers and chemical control of pests.

Although much energy has been devoted to understanding consumptive uses of water much remains to be learned. A thorough and comprehensive understanding of the determinants of various consumptive uses will be needed if sophisticated water management schemes are to be developed. Agricultural uses are both large and important. Yet, understanding of agricultural water use is far from comprehensive. Agriculture water is frequently seen as the supplier of last resort because it accounts for such a large share of total consumptive use. It seems likely that food production will need to be increased sharply worldwide if a growing global population is to be fed. This suggests that it will be critically important to learn how to manage agricultural water supplies efficiently so that they yield optimum levels of production of food and fiber.

NONCONSUMPTIVE USES OF WATER

There are a number of water uses that by their nature do not render the water unfit for subsequent use. Among these are navigation, hydroelectric power generation, thermal cooling (within limits), and environmental uses. Of these, environmental uses have assumed particular importance because it is now understood that water development and use practices of the past have severely constrained and reduced the water available for environmental uses. Environmental uses of water provide both amenity values and service values. Amenity values include water-based recreational values, scenic values, and option values that accrue to individuals who appreciate the presence of pleasing water-based environments even if they never expect to see or use them. Service values include the benefits of environmental services provided by aquatic ecosystems such as air and water purification and the inherent environmental stability associated with diverse ecosystems. In instances in which the capacity of water-based environments to provide ecosystem services is impaired, it is very expensive to provide those services artificially in lieu of the environmental services, and the resulting in lieu services are never as effective and efficient as they are when provided by the environment.

There is very little understanding of the role of aquatic ecosystems and the role of water in supporting environmental services, with the result that there is a compelling need to understand such ecosystems in a broad systems context. Additional research will be required to understand the determinants of water requirements for the maintenance of aquatic and riparian ecosystems in order to preserve their capacity to provide wildlife habitat, flood control and assimilation, and dilution of contaminants. There is much interest in restoring the flows of some of America's major rivers such as the Colorado and Missouri to conditions that mimic their undeveloped states. If this is to occur it will be necessary to develop a systematic understanding of the relations between biological, hydrological, and geological factors.

It will also be necessary to develop a better understanding of the relationship between land and water resources if water is to be effectively managed on a watershed basis. Efforts to develop the knowledge needed for innovations in watershed management have been hindered by the tendency to study terrestrial and aquatic ecosystems independently of each other. Additional research will be needed on a whole range of issues that bear on the protection of species diversity in aquatic habitats. Only by preserving species diversity will it be possible to maintain environmental stability and ecosystem health in aquatic ecosystems.

The discussion in both this and the preceding sections provide a sample of the kinds of knowledge that must be acquired if the United States is to address successfully the water management challenges of the next several decades. Some of these issues may be usefully addressed by groups of scientists in a collaborative way that would yield richer results than if they were addressed independently. There are, of course, hundreds of such issues and those identified here are simply meant to be suggestive.

CONCLUSIONS

This brief overview of the water resources of the United States and the various problems and challenges that attend to the management of those resources is intended to paint a broad picture. The situation in the semiarid portion of the country was emphasized because the physical circumstances and problems of management there are more akin to those encountered in Iran. Several general conclusions emerge from the overview as follows:

- Water scarcity will persist and intensify as population and the economy grow.
- Learning how to manage water scarcity with reasonable efficiency will be a continuing challenge.

- The management of water quality will continue to be challenging. Control of nonpoint source pollution will be particularly important as will the management of new chemicals and technologies to avoid further contamination. The contamination of groundwater will be a growing problem and in situ techniques will offer the best and least costly methods of clean-up.
- The need to manage groundwater sustainably will continue to be important in many regions of the country.
- The management of salinity will pose a continuing challenge in many areas where irrigated agriculture is practiced.
- Additional research and technical innovation will be needed if consumptive and nonconsumptive uses are to be fully understood and managed effectively.

Overview of Water Management in Iran

Reza Ardakanian

ABSTRACT

At the present time, improvements in water resources management are being sought and implemented as part of the course of socioeconomic changes in Iran, including transformation in the structure of the national economic system and demand-supply mechanism for water. Changes also are being prompted by emerging issues resulting from regional and global water crises.

Studies indicate that the present system is entering a new stage, with widespread economic and environmental consequences arising from its progression over the past 70 years. The current condition has revealed the necessity for adopting coherent, farsighted, and comprehensive plans and actions.

The paper describes the conditions and necessary infrastructure for transitioning from the present stage to a new one. First, the past and present situation of water resources management in Iran is presented. Then, the assessment of future changes, including long-term policies and a related action plan are presented. Finally, a list of plans that were executed in the past and some of their results are presented, with special emphasis on two important factors of water conservation and water reuse.

WATER MANAGEMENT AND SOCIO-HISTORICAL EVOLUTION

The present system of water resources management in Iran began to evolve about 70 years ago under certain historical and social conditions. The general progression of this evolution can be summarized as follows:

In the last century (since 1900) the population of the country has increased about six-fold. The population growth rate, which was less than 0.6 percent in the beginning of this period, reached the rate of 3.19 percent in the decade from 1976-1986. Fortunately, it has considerably decreased once again in the last decade. The major changes in population growth rate, resulting from reduction of mortality and increase of natural growth rate, occurred in the 1960s and afterward. Part of the population growth of the last decade has been due to immigration of Afghan refugees. Between 1960 and 1996, about 37 million people (about 60 percent of the existing population) were added to the country's population.

In the period from 1961-2000, the urban population increased by about 31.7 million and the rural population increased by 11 million. In 1956, there were only three cities with a population over 250,000 in Iran, while in 2000 the number of cities with a population of over one million reached seven.

The direct impact of population growth on the water resources management of the country was an increased need for potable water in population centers. Indirect impacts were increased demand for agricultural products, development of irrigated lands, and the need for job opportunities and more income, especially in the agricultural sector.

The impacts of rapid urbanization included an increased domestic use of water, especially for hygienic purposes, and the emergence of new water needs due to the expansion of cities and improvements in living standards. Under such conditions, new responsibilities have been created for water resources management, of which the most important are the increased importance of protecting population centers against drought and flood, and the ever increasing importance of water treatment to provide hygienic water, as well as collection and sound disposal of wastewater and drainage water.

Along with changes that have been described in the areas of population increase, urban development, exploitation of water resources, qualitative and quantitative limitations for these resources, economization and protection of water resources, and protection of the aquatic environment against water pollution have gained importance.

Evolution in the Political and Administration System

As political and administrative institutions expanded and government became more centralized, especially after the 1960s, the role of planning and budgeting in the fate of the country became more important. After the Islamic Revolution of 1979, related social and political changes revealed the need for reform in political and social systems through supporting parliament, encouraging public participation, and privatizing and liberalizing the economy. The needed changes encompass the water management system of the country, including several water resource development projects under implementation, or nearing implementation.

Fundamental Changes in the Economic System

Existing information indicates that the increase in national economic production before the late 1950s was based on provision of minimum subsistence in society. Since the last years of the Third Development Plan of the country (1966), economic surplus has been taken advantage of for development planning of the country.

During the fundamental changes in the economic system of the country, investment in the development of water resources (both by governmental and private sectors) has considerably increased, and the system of water resources utilization has undergone drastic changes.

Changes in the Utilization System of Water Resources

The appearance of major changes in the water resources utilization system of Iran dates back to the early 1960s. Since then, about 58 big reservoir dams have been constructed and have become operational. The volume of their regulated water is more than 30 billion cubic meters (bcm). Over-exploitation of groundwater has taken place by substituting deep and semi-deep wells in place of qanats. The volume of exploited groundwater has increased by 2.7 times.

In total, these changes have resulted in an increase of water exploitation (both from surface water resources and groundwater) from 40 bcm to 90 bcm in 1996 (2.25 times). These changes have expanded the contribution of secure water from 65 percent to 82 percent, and the exploitation of surplus water from watersheds and aquifers to about 5 bcm.

Due to the development of water resources and an increased distance between water supply centers and points of use, transmission systems and technology expanded considerably and in complex ways. In urban water supply systems, transmission pipelines, tunnels, pumping stations, and physical treatment (with increased contribution of surface water in securing urban water) have become more important. Along with the government's increased role in implementation and operation of projects on water security, water supply for cities, irrigation, and drainage networks; the contribution of the governmental sector to the utilization system has also increased.

Changes in the Water Resources Management System

The trend of changes in water resources management systems can be divided into three stages as follows:

- Commencement stage (1927-1963)
- Shaping stage (1963-1979)
- Changing stage and transfer to new stage (1979 until present).

It should be mentioned that the water resources management system has become primarily technology oriented (construction oriented) since the 1960s.

The policies on institutional reforms for privatization of more liabilities and mitigating obstructive regulations and laws have been on the agenda of the Cabinet and the Islamic Consultation Assembly (Parliament) since the beginning of the First 5-Year Development Plan of the Islamic Republic of Iran. Along with these policies and with respect to existing limitations for water related services, the water management of the country has put in force several actions, and many others are being implemented. Among the most important actions under implementation is the establishment of independent urban Water and Wastewater Companies. Another important action is to commit the maintenance and operation of irrigation and drainage networks to farmers and to support them by providing financial facilities. Part of the investments pertaining to urban sewage, irrigation, and drainage projects is based on the Constitution.

WATER SUPPLY RESOURCES

The main source of water resources throughout the country is annual precipitation. According to studies carried out for formulation of the Water Comprehensive Plan, the main characteristics of annual precipitation and its conversion to water resources are as follows:

- Average annual precipitation 417 bcm
- Average annual evaporation & transpiration 299 bcm
- Surface currents 92 bcm
- Direct seepage to alluvial aquifers 25 bcm

According to the above figures:

- About 72 percent of precipitation is not accessible due to evaporation and transpiration,
- About 22 percent of precipitation flows as surface water resources,
- About 6 percent of precipitation within the borders of the country is used for direct recharge of alluvial aquifers.

Consequently, about 117 bcm of water is directly and potentially accessible by people through precipitation (internal renewable resources) each year.

In addition to water resources gained through precipitation within the limits of the country, about 13 bcm of surface flow enters the country across its borders. When this flow is combined with the surface flow with internal origins, the total figure of surface water resources of the country increases to about 105 bcm. Of this amount, about 13 percent (13 bcm) is used for recharge of alluvial aquifers.

Accordingly, annually about 130 bcm of water is accessible for people through precipitation and inflow currents across borders (total renewable resources).

In addition to naturally processed water resources, about 29 bcm of exploited and consumed water from surface and groundwater resources appears again as exploitable surface water or penetrates to alluvial aquifers as reservoirs. Correspondingly, the total water resources of the country, including such water exchange processes, increase to about 159 bcm. Out of this, 82 percent (130 bcm) are renewable sources, and 18 percent (29 bcm) are return waters that are discharged into surface and groundwater resources and are included in the calculation of total water resources. As annual changes in quantity and quality of consumption patterns take place, this section of water resources also changes quantitatively and qualitatively.

In the year 2000, about 43 bcm of surface water resources, including regulated flows, were exploited by reservoir dams, pumping stations, small scale water supply projects, or traditional stream systems.

According to present conditions, the amount of groundwater is estimated to be 47 bcm. In the year 2000, total exploitation of groundwater resources was 90 bcm, which is 70 percent of renewable water resources and 57 percent of total water resources (renewable, water exchange, and return flow).

WATER USES BY SECTOR

The greatest amount of water use (83.5 bcm or 92.8 percent) is by the agricultural sector. Of this amount, about 50 percent is exploited from surface water resources and another 50 percent from groundwater.

Exploitation of water resources by the mining sector is about 1.1 bcm or 1.2 percent of total use. About 54 percent of water utilization in this sector is from groundwater resources and the remaining amount is from surface water.

Withdrawal of water by the urban and rural water supply sectors is about 5.4 bcm, or 6 percent of total water exploitation of the country, of which about 68 percent is from groundwater resources and the remaining 32 percent from surface water.

WATER AND ECONOMIC ACTIVITIES

Water resources management services and related activities are linked with economic structure as well as the trend of economic development. Regarding the existing economic structure of the country, the main points concerning the importance of water follow.

Those who manage and exploit water resources at different levels have made high investments for improving and regulating related services. Governmental investments have mainly occurred at a nationwide level, aimed at control and distribution of surface water related to consumption, energy generation, and flood control. Private investments have mainly occurred as disorganized and local efforts aimed at exploitation of groundwater or pumping of water from an adjacent river. Total gross investment until now, based on the fixed prices of year 2000, is estimated to be 100 trillion Rials (U.S. $12.5 billion), of which about 40 percent has been invested by the private sector.

Gross capital investments have been about 1.2 percent of gross domestic product (GDP) and 5.8 percent of gross national investment. Final cost for water supply from reservoir dams based on fixed prices compared to the 1960s is eight times as much, and the cost of securing water from wells has been tripled.

The annual economic value of water and related services (provision and distribution) for agricultural and urban use as well as energy generation, based on 1996 prices, is estimated to be 9.7 trillion Rials (U.S. $1.2 billion). This amount, when combined with the value of water supply in rural areas, water usage by the industrial and mining sectors, recreational utilization, and aquaculture activities that total 10 percent of the above figure, is estimated as 10.7 trillion Rials (U.S. $1.3 billion) per year. This figure represents 8 percent of GDP without including crude oil prices and 6.7 percent of GDP if the oil sector is included.

WATER AND FOOD SECURITY

Currently, the contribution of irrigated lands to production of cereals is about 69 percent and to production of other products (horticulture and orchards), about 90 to 100 percent. Different methods of securing and exploiting water resources provide different contributions to securing food supplies. With an increased contribution of assured water resources toward total agricultural water resources, the condition of food security has improved to some extent. But other contributing factors, such as productivity of water resources and irrigation standards, have not so improved. Moreover, it remains necessary to strengthen drought management institutions and improve food security during periods of drought in the country.

STRUCTURE OF WATER MANAGEMENT

At present, the main institution for water resources management is based in the Ministry of Energy, and its main components are as follows:

- Deputy Minister for Water Affairs (Iran Water Resources Management Organization),
- Regional water companies,
- Water and Wastewater Engineering Company (nationwide), and
- Provincial Water and Wastewater companies, also in important cities (30 companies).

Furthermore, about 124 consulting firms and 216 construction companies support the above sections.

In the framework of sectional planning, the effective water resources management system has an influence on different social and economic sectors, and reciprocally these sectors leave their impression on the water management system as well. Among those ministries and organizations whose activities have notable impacts on water management systems are Ministries of Agriculture, Industries and Mines, Housing and Urban Development, Interior, Health, Roads and Transportation, and finally, Department of the Environment. For collaboration and coordination between the above ministries and organizations, the Supreme Council of Water has recently been established. This Council is presided over by the President of the Republic. All related organizations and ministries, as well as parliamentary representatives, are members of this Council. It is worth mentioning that in the parliament, different committees on water, agriculture, natural resources, budgeting, and development, supervise management activities all across Iran.

POSSIBILITIES AND OPPORTUNITIES FOR WATER RESOURCES MANAGEMENT

Possibilities for improvements of water resources management exist because of Iran's vast areas of cultivable lands, big rivers, suitable sites for construction of dams in Zagross and Alborz mountain ranges, extensive aquifers, and suitable climatic conditions for cultivation of different plants. In addition to existing exploitation of the water resources of the country, there is some capacity for physical development of up to 30 bcm of water resources for consumptive uses and up to 50 bcm for energy production, while still observing all economic, social, and environmental limitations.

Cultural support for developments in water resources management includes traditions and social institutions that have adapted over time to different geographical conditions, especially in arid and semiarid regions. The best adapted of such institutions allow for effective water resource utilization.

Political conditions conducive to developing water resources management include extensive citizen participation in public affairs, strengthening of the parliamentary system, creation of nongovernmental organizations (NGOs), support of local management, and suitable political background for developing the water and agriculture sector. Existing installations and equipment include large and small reservoir dams, extensive irrigation and drainage networks, water transmission pipelines and pumping stations, treatment plants and water reservoirs, urban water distribution networks, and other facilities. Institutional capacity includes the possibility for experienced national experts to provide consulting and construction services in order to reduce foreign exchange expenses to a considerable extent.

Allocation of more water for urban and industrial uses, expansion of non-consumptive uses of water resources (hydroelectric generation and recreational use of water bodies), and cultivation of profitable agricultural products has led to increased economic value of water, enabling more capacity building for sustainable and economic use of water resources. Reducing limitations and increasing competition will accelerate this process.

International Contributions

Due to endeavors by international organizations to water resource management in the early 1990s, the policies for attaining sustainable development have entered a new stage. Although this attention was given to natural resources in general, and was not confined to water resources, the water-based problems received particular attention because of their complexity and severity. New efforts to overcome previous obstacles to implementation of the action plan (since a 1977 International Conference on Water) have led to more international conferences and seminars planned for the near future. Such international organizations have had success in confronting regional water conflicts and managing international water basins. International cooperation in these areas can mitigate the problems considerably. Hence, enhancement of international relations can provide opportunities for improvements in water resource management of the country by making the best use of existing knowledge of other countries and fostering negotiation on common issues. One of the advantages of water resources management in Iran is low dependency on international water resources. Only about 7 to 8 percent of water resources of the country are secured from international or common borders.

CHALLENGES OF WATER RESOURCE MANAGEMENT

Renewable water resources of the country are estimated to be about 130 bcm. Because of rapid population growth, per capita water resources have steadily decreased and will continue to decrease in the future. Geographic distribution of water resources of the country has not been consistent with geographic distribution of population, especially in the last two decades. Hence, there is a growing need for more water transfers in and between basins.

The transition from an agricultural economy and renewal of agricultural structure is not yet complete. Land ownership and agricultural activities are still going through transition, and agricultural development still happens mainly through expansion of irrigated lands.

In spite of previous endeavors, it is necessary to strengthen the following aspects of water resources management:

- policy formulation,

- laws, regulations, criteria, and standards,
- organizational improvement (coordination, cooperation, different specialization, and decision making processes),
- water allocation system,
- personnel planning and management,
- financial and economic management,
- information systems and data banks, and
- technological research and development.

Evidence indicates natural and human-made occurrences of destructive floods all over the country. Prevention and mitigation of adverse impacts and losses resulting from such destructive natural disasters are of utmost importance. Normally, cities are more vulnerable to such impacts. Coordination of responsible organizations plays a vital role in such prevention and mitigation.

Periodic occurrence of drought is always possible at regional and national levels. Drought has a substantial impact on domestic food production and its contribution to meeting fundamental food needs of Iran's population. Thus, forecasting and adopting appropriate measures for dealing with drought by water management authorities and other organizations is very important from a national security point of view.

The social problems related to the management of water resources are aggravated when impacts of water-related actions are adverse. Measurement of the impacts and remedies for them have key roles in upgrading validity and accuracy of decisions. Environmental criteria and standards must be observed when implementing activities of water resources management. The impacts in large-scale water resource projects and in exploitation of groundwater should be paid due attention.

DETERMINATION OF ESSENTIAL IMPROVEMENTS FOR THE FUTURE

The existing system of management and exploitation of water resources of Iran has been shaped by the events of the 1960s. Conditions and events since then have increased the importance of national management of water in macroeconomic planning of the country. The increased need for national planning and expansion of water resources management will continue into the future.

Conditions and trends that support the need for national level water resources management can be divided into three groups: those past and present conditions that have impacted water resource management and will continue to do so, recent developments whose affects on water resource management can be expected to accelerate into the future, and new trends or conditions that may be expected to have an impact in the future.

Past and Present Factors Affecting Water Resources Management

The following are conditions and trends that in the past and present have impacted the water resources management system and will continue to do so:

- Population growth, rapid urbanization, and increase of per capita consumption of urban water have changed the proportion of water utilization in population centers.
- Growth of industry and rapid industrialization of the country have caused an increasing proportion of industrial water consumption.
- The growing importance of planning and future forecasting, especially in long term planning, will promote the role of water resources management in development plans.

Recent and Growing Developments Affecting Water Resources Management

Recent trends that will be accelerated in the future are as follows:

- Increase in nonconsumptive uses of water resources for energy generation, aquaculture (cold water fish), and recreational purposes.
- Greater sensitivity and awareness about pollution of water resources and its economic and environmental consequences.
- Greater sensitivity and awareness of adverse environmental impacts of water resources development projects.
- Increase in the cost of securing additional water.
- Prevention of unwise utilization of groundwater resources.
- Increase in relative contribution of small-scale projects, with more participation of local water users.
- Increase of the contribution of water substitute management and water renewal installations.
- Reduction of urban and agricultural water loss, especially in transmission and distribution phases.
- Collection and sound disposal of urban sewage.

Potential Future Factors Affecting Water Resources Management

The following trends may play an increasing role in the future of water resources management:

- Increased pressure to develop surface water resources as compared with groundwater.
- Collection and sound disposal of industrial wastewater.

- Treatment and reuse of urban and industrial wastewater.
- Participation of local residents, users, and stakeholders in financing large-scale water resource development projects.
- Reduction of water loss from farms.
- Increased importance of mobilizing financial resources, and of pricing projects and implementation programs.
- Increased role of national and regional water management in foreign policies, national spatial strategy plans, national physical plans, and regional planning simultaneously with increasing importance of common water basins and export of water, the role of water in national security and food security, and the necessity of adopting basic solutions in critical water basins.

It is anticipated that these trends and consequent changes will continue in such a way that national management of water resources will enter a new stage with wide socioeconomic and environmental dimensions. As this stage commences, new duties will confront the water managers of the country, and they may need to gain additional qualifications. Two major actions needed are as follows:

- Remove or mitigate inadequacies of existing infrastructure and water management systems, and
- build capacity in water management, especially regarding demand management.

For a more specific comparison of the main changes in the aspects of water resources management that have been discussed in this section of the paper, some selected indexes are presented in Table 1, comparing the base year (2000) and a target year (2021). The annual volume of water available for consumptive uses will increase from 97 bcm to 120 bcm (according to projections). In total, about 23 bcm of new capacity is expected to occur. About 92 percent of total capacity of available water will be used at that time. The volume of exploited water will be increased by 138 percent compared to the base year. A part of the additional water utilization will be derived from development of existing unused capacities and the other part will be derived from development of new capacities. Current unused capacities pertain to large reservoir dams. At present, about 8.5 bcm out of a total capacity of 25.4 bcm is not exploited.

In order to meet fundamental objectives and determine the main course of movement from present conditions to future ones, an action plan has been formulated. This plan consists of 5 general policies and 35 adaptable strategies (Table 2). Under this action plan framework, an independent execution program can be prepared for each strategy, and the relationships between all the strategies can be derived through determination of general objectives. The main objective is efficient and equitable development and utilization of water resources of the

TABLE 1 Baseline (year 2000) and Projected (year 2021) Characteristics of Water Resources Management in Iran

Indicator	Unit	2000	2021 (selected scenario)	Ratio (%)	Percent change
Total volume of exploited water	bcm	97	120	124	+24
Share of water resources, by source					
—Groundwater	%	52	42	81	–19
—Surface water	%	48	55	114	+15
—Recycled (domestic, industrial)	%	—	3	—	
Share of consumption, by sector					
—Agriculture and aquaculture	%	94	86	92	–9
—Urban & Rural	%	6	7	117	+17
—Industry & Mine	%	1.2	3	250	+150
Water loss, by sector					
—Agriculture	%	64	60	94	–6
—Urban	%	27	10	37	–63
Volume of return flow	bcm/yr	29	40	138	+38
Effluents and Wastewater	bcm/yr	4.5	8	178	+78
—Urban	bcm/yr	3.7	5.5	149	+49
—Industrial	bcm/yr	0.8	2.5	312	+213
Investment					
—Total gross investment	10^{12} Rls	41	262	635	+539
—Contribution of private sector	10^{12} Rls	40	32	80	–20
Importance in national economy (NE)					
—Contribution of water investment from GDP	%	1.2	2.6	217	+117
—Contribution of water value and related services in NE	%	7.5	9.8	131	+31
Contribution of capital return of expenses of governmental projects					
—Urban water	%	22	50	227	+127
—Agriculture water	%	6	23	383	+283
Economic revenue of water in different sectors (average)	Rls/cm	1614	5018	311	+211
Economic revenue of water in farming subsector					
—Productivity of agricultural water	kg/ m^3	0.6	1.1	183	+83
Water and food security					
—Role of water in the production of cereals	%	69	73	106	+6
—Role of water in the production of other yields	%	90-100	90-100	—	

TABLE 2 Long-Term Policies and Strategies for National Management of Water Resources of Iran

Policies	Strategies
A. Establish a comprehensive water management system that incorporates natural elements of the total water cycle as part of principles of sustainable development, and makes use of a "national spatial strategy plan" based on natural water basins of the country.	1. Establish a management system for water resources of the country based on integrating and observing continuous elements of the water cycle. 2. Strengthen the main water resources management institution with special emphasis on establishing a comprehensive system for water allocation. 3. Establish suitable multisectional coordination institutions at national and local levels as well as in water basins. 4. Decentralize and develop water resources management abilities at different levels (capacity building). 5. Integrate the plans for development, exploitation, and protection of water resources with other national and regional plans. 6. Prioritize the role of water resources management of the country in planning systems, so that goals, limitations, possibilities, and strategies governing water resources management are reflected in the process of formulating national and regional plans. 7. Promote financial management and capital mobilization in accordance with changes in water resources management.
B. Improve productivity by promoting the economic, political, and national security values of making improvements in water utilization, supply, protection, and consumption.	8. Promote public awareness, improvement and transparency of tariffs, and utilization of technical, economic, and managerial instruments. 9. Make better use of qualified personnel and natural potential as well as existing social and local institutions. 10. Anticipate how to mobilize necessary financial resources compatible with value added from development of water resources (appropriate contribution from GDP). 11. Balance the needs for efficiency and social equitability. 12. Promote the advantages and importance of water resources management in food security and national security. 13. Promote productivity capacity in the process of water management cycle and effective factors in economic value of water. 14. Encourage research, education, and propagation of information.
C, Increase the rate and quantity of water utilization, and minimize any means of water loss.	15. Develop water resources under national plans and under comprehensive watershed-based plans. 16. Prepare, compile, and execute comprehensive research plans for reducing water loss. 17. Prepare and compile needed research programs. 18. Develop skills and technical know-how to expand upon indigenous skills and optimize prices. 19. Create needed capacities for developing and mobilizing nongovernmental and foreign financial resources.

continued

TABLE 2 Continued

Policies	Strategies
D. Compile a comprehensive plan for allocating resources among implementation of projects on dams, watershed management, aquifer management, irrigation networks, water quality, drought and famine relief, flood prevention, recycling and utilization of unconventional water resources, development of technological competency, and public participation in water development.	20. Coordinate the timing and scheduling of preparation and execution of complementary water projects. 21. Strengthen water quality management in water resources management systems. 22. Strengthen multisectional coordination mechanisms for enforcement of water quality management policies. 23. Enforce organizational changes for internal and external decentralization, with emphasis on public participation. 24. Make the best use of administrative, financial, and educational mechanisms for encouraging public participation. 25. Prepare and implement comprehensive programs for flood management at a watershed level. 26. Develop necessary programs for drought relief and crisis management. 27. Prepare master plans and implement projects for water recycling and utilization of nontraditional water resources. 28. Promote the development of knowledge and information systems to expand technological know-how. 29. Support industrial plans needed by water resource management and enforce them. 30. Prepare and compile engineering master plans for rivers and riverbanks. 31. Encourage and develop all means of collecting, processing, and disseminating baseline information.
E. Control and manage water bodies whose flow naturally exits the country, and prioritize utilization of joint (international) water resources.	32. Prioritize the development of comprehensive plans for controlling border waters, and adopt programs for securing the necessary financing. 33. Encourage regional cooperation and exchange of information. 34. Establish needed structures for coordination in policy and decision making. 35. Prioritize the establishment of qualitative and quantitative data collection and processing systems for information related to water resources along common borders.

country in accordance with the socioeconomic and environmental needs of present and future generations.

The five general principles (policies) prepared for this action plan and approved by the Expediency Council, are as follows:

1. Establish a comprehensive water management system that incorporates natural elements of the total water cycle as part of principles of sustainable

development, and makes use of a "national spatial strategy plan" based on natural water basins of the country.

2. Improve productivity by promoting the economic, political, and national security values of making improvements in water utilization, supply, protection, and consumption.

3. Increase the rate and quantity of water utilization, and minimize any means of water loss.

4. Compile a comprehensive plan for allocating resources among implementation of projects on dams, watershed management, aquifer management, irrigation networks, water quality, drought and famine relief, flood prevention, recycling and utilization of unconventional water resources, development of technological competency, and public participation in water development.

5. Control and manage water bodies whose flow naturally exits the country, and prioritize utilization of joint (international) water resources.

ACCOMPLISHED ACTIONS

In view of the fact that approval of long-term policies coincided with the commencement of the Third Five-Year Development Plan of the country, there was much effort during the first two years of this plan to carry out these policies. The most important achievements are as follows:

- The allocated budget for the water sector in the Third Plan has increased by 300 percent (tripled) compared to the Second Plan.
- Financial resources for the water sector were diversified, including making use of foreign investment in some important projects.
- Two ministries of Agriculture and Jihad Sazandegi (Crusade for Construction) were merged for upgrading agricultural and natural resources activities. This will play an important role in sectional coordination.
- The Supreme Council of Water was established, presided over by the President of the Republic and having membership composed of high-level experts and authorities.
- A Water Comprehensive Plan projecting to the year 2021 was prepared, containing 54 volumes.
- The preparation and compilation of the Water Comprehensive Plan of the country was put in the agenda of the Cabinet regarding present and future changes.
- In order to strengthen regional cooperation, the Regional Center on Urban Water Management (RCUWM-Tehran) was established in 2001 with collaboration of the United Nations Educational, Scientific, and Cultural Organization (UNESCO). Duties of the RCUWM-Tehran are to promote the transfer of technical knowledge and experience and to promote awareness and capacities in all aspects of urban water management. These duties are aimed at promoting sustainable development and making use of the results of regional activities for the

relative welfare of the population of countries of the region. It is worth mentioning that for commencement of the activities of RCUWM-Tehran and planning for the future, recently a contract has been established with UNESCO's Institute for Water Education (IHE).

- For supporting research activities in the water sector, a Water Research Institute has been established. This institute will concentrate on assessing the water resources of the country and enhancing the quality and quantity of related data and information.
- Extensive national-level programs have begun to be implemented to optimize consumption patterns in the agricultural sector. It is anticipated that with implementation of these programs, major steps will be taken for managing demands in agriculture and promoting water value.
- A new system of water allocation was developed and put in force to manage existing demands.
- Attention was paid to wastewater treatment and water reuse regarding the quality management system.

Some quantitative goals that have been accomplished are as follows:

- With construction of 12 new reservoir dams, about 3,700 bcm of additional regulated water was supplied for use by different sectors.
- About 140,000 hectares of irrigation and drainage network was constructed.
- With implementation of small scale water projects, more than 352 million cubic meters (Mm^3) were secured as regional contribution to water resources development projects.

TREATMENT TECHNOLOGIES

Large Scale Systems

Stephen M. Lacy

ABSTRACT

The arid Southwest of the United States has a very similar climate to portions of Iran. With little rainfall, communities must share freshwater sources from rivers and aquifer systems with agriculture. The scarcity of freshwater sources is driving decisions to look at wastewater effluent as a resource. Recent technological advances borrowed from the water industry are opening up more opportunities for the safe reuse of effluent. This paper will look at the treatment standards commonly set for several categories of reuse and the treatment technologies that are being employed to obtain differing levels of effluent quality to meet the needs of the reuse opportunities. It will also look at new emerging technologies in the area of microfiltration and how it is being applied in reuse. Finally it will look at the treatment technologies being incorporated for indirect potable reuse applications.

INTRODUCTION

In the past, reuse of wastewater effluent was typically limited to irrigation of agricultural pasture lands and turf areas with limited public access. As freshwater sources become more limited, other applications for reuse are being employed that result in greater contact between the effluent and the public and food crops. This has invited ever-increasing regulation and restriction, thus requiring more sophisticated treatment processes to be incorporated into the typical treatment train of a wastewater facility.

Wastewater reuse provides a drought-proof resource for a community that automatically increases as population growth occurs. Reuse also provides for some immediate reductions in water diversions by replacing existing demands (such as turf irrigation at parks) with nonpotable irrigation water. This water resource offers strategic benefits to the community in terms of increased sustainability. However, because most nonpotable uses (such as irrigation) are limited and the demand diminishes in the winter months, many communities look at other options for use of the resource rather than simple discharge. Figure 1 shows wastewater production levels and a typical variation in annual demand for nonpotable irrigation uses.

During periods of low irrigation demand, wastewater reuse can also be accomplished by indirect potable reuse. Indirect potable reuse occurs where highly treated reclaimed water is introduced to a surface water or groundwater system that is ultimately used as a potable water supply. In an indirect system, the reclaimed water is blended with the natural system, with a significant delay (12 months or more) between the point of reclaimed water discharge and the withdrawal into the potable water system. Dilution of the reclaimed water with natural waters results in only a portion of the water being withdrawn for potable use originating from the reclaimed water. There is a very significant direct benefit to the water resource by reducing groundwater depletions and pumping effects through aquifer recharge. In order to maximize beneficial reuse, a range of water reuse alternatives and wastewater treatment technologies is needed.

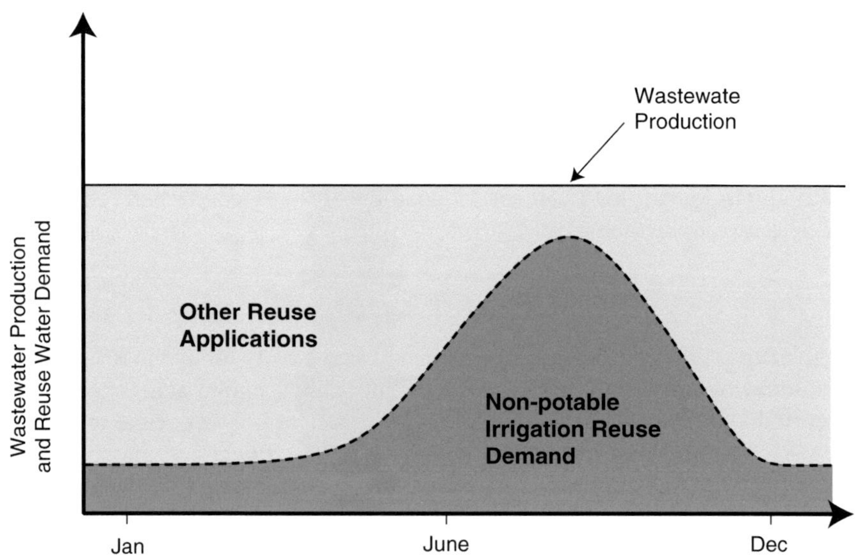

FIGURE 1 Typical annual irrigation demand pattern.

COMPARISON OF WATER QUALITY STANDARDS FOR NONPOTABLE REUSE

Water quality standards for nonpotable uses have been established by many states within the United States and continue to be revised as a result of developing technologies in wastewater treatment and a better understanding of the health effects. The focus of these standards is to provide policy direction and regulation of reuse applications that are protective of public health. California's Title 22 Standards have been in use since 1978, and the latest revisions were adopted in December 2000.

The level of public exposure and contact typically categorizes the allowed effluent irrigation uses. Prescriptive treatment standards and water quality limits are listed for the various categories of use and are widely accepted to be protective of public health. There are many examples throughout the western United States where reclaimed water has been safely used for many years to irrigate turf (golf courses, parks, and recreational sport fields). As communities look to expand beneficial use of reclaimed water, it is expected that many more nonpotable reuse applications will be implemented to preserve and extend potable supplies. The following discussion summarizes the typical reuse categories identified in regulations: agricultural reuse, urban reuse, and industrial reuse.

Restricted Reuse for Commercially Processed Food Crops and for Non-Food Crops

This category describes how effluent may be used to irrigate commercially grown food crops that must undergo commercial pathogen-destroying processing before being consumed. It also includes irrigation of non-food crops, such as hay and seed crops, or for food crops where the reclaimed water does not come into contact with the edible portion of the crop (orchards or vineyards). California standards require secondary treatment (≤30 mg/1 biochemical oxygen demand [BOD] and total suspended solids [TSS]). Newer regulations additionally require disinfection to ≤200 fecal coliforms/100 milliliters (ml) (median for prior seven samples) with no single sample exceeding 800 fecal coliforms/100 ml.

Common activated sludge treatment with simple disinfection can meet these criteria. If the disinfection limit is higher, such as 1,000 MPN/100 ml, a lagoon system consisting of facultative lagoons followed by maturation ponds can meet secondary standards with long detention times.

This category does not include pasture for animals producing milk for human consumption. If the reclaimed water is to be used for irrigation of pastures for milking animals, the disinfection must be increased to medium-level. This would require additional disinfection facilities to consistently disinfect the reclaimed water to a fecal coliform bacteria concentration of less than 23 fecal coliforms/100 ml (median for the prior seven samples), with no single sample exceeding 92 fecal coliforms/100 ml.

Restricted Reuse for Non-Commercial Food Crops

This category describes how reclaimed water may be used for surface or spray irrigation of food crops that can be consumed raw or that are not commercially processed. California requires that reclaimed water receive secondary and tertiary treatment followed by high-level disinfection. This standard is identical to the standard for urban irrigation use with unrestricted access described in the following section. The organic content must be reduced to less than 10 mg/l of BOD, and the solids to less than 2 nephelometric turbidity units (NTU) (5 mg/l TSS). Fecal coliform bacteria must be reduced to less than 2.2 MPN/100 mg/l. Common activated sludge treatment followed by sand filtration will comply with the organic and solids limits. Extensive disinfection facilities are required to consistently meet the coliform levels.

Urban Irrigation Use—Restricted Access

This application typically involves minimal public exposure to the irrigation water and is used where public access is prohibited, restricted or infrequent. Examples include freeway landscapes, cemeteries, sod farms, silviculture sites, and potentially golf courses where public access is restricted. California standards allow use of disinfected secondary treated (≤30 mg/l BOD and TSS) effluent. The California standard has set a medium level of disinfection for this use category, ≤23 total coliform/100 milliliters (ml) based on the last seven samples. It is expected that most nonpotable uses identified in a community will not qualify as restricted access. Common activated sludge treatment with simple disinfection can meet these criteria.

Urban Irrigation Use—Unrestricted Access

This category describes how reclaimed water may be used to irrigate landscaping where people walk, play, or otherwise spend time. Examples of these uses are playgrounds, schoolyards, sports fields, public golf courses, and parks. The reclaimed water must meet the highest standards for reuse to protect the public from exposure to pathogens found in domestic wastewater. California standards require secondary treatment with filtration (tertiary treatment) and a high level of disinfection. The standards additionally require coagulation prior to filtration and that the filter effluent turbidity does not exceed 2 NTU and the filter influent turbidity does not exceed 5 NTU for more than 15 minutes and never exceeds 10 NTU. The disinfection standard requires that the effluent achieve ≤ 2.2 total coliforms/100 ml (median for the prior seven samples), with no single sample exceeding 23 total coliforms/100 ml. It is expected that most nonpotable uses in a community will be required to meet this treatment and disinfection standard.

The organic content must be reduced to less than 10 mg/l of BOD and the solids to less than 2 NTU (5 mg/l TSS). Fecal coliform bacteria must be reduced to less than 2.2 MPN/100 mg/l. Common activated sludge treatment followed by sand filtration will comply with the organic and solids limits. Extensive disinfection facilities are required to consistently meet the coliform levels. Coagulation will be required as part of the filtration treatment to achieve consistently low filter turbidities. Flocculation may be required as well.

Industrial Reuse

The use of reclaimed water in industrial applications must comply with the quality levels set above for the level of contact by the public in urban irrigation applications. Additional treatment may be required to meet water quality for the industrial process for direct use or to meet a prescribed pretreatment level. In most cases, the unrestricted access irrigation use category is of adequate quality to meet larger industrial uses such as cooling water and may be of adequate quality for industrial process water applications.

TREATMENT TECHNOLOGIES TO MEET NONPOTABLE WATER REUSE STANDARDS

Different levels of treatment are required to implement different reuse strategies. They range from simple upgrades to the existing treatment facilities for irrigation of agricultural and urban areas, to multiple barrier treatment for indirect potable reuse. Treatment technologies for indirect potable reuse are discussed later in the paper.

Common to each strategy is the need for preliminary and secondary treatment prior to any advanced treatment. Preliminary treatment would include screening and grit removal. Secondary treatment would consist of a biological treatment process such as activated sludge or trickling filters. A lagoon system could be utilized in a restricted use agricultural application; however, the low quality effluent and high algae content in a lagoon effluent makes the water unacceptable for further high level treatment. Primary clarification could be included as a unit process prior to secondary treatment in larger plants. Figure 2 shows a typical configuration of a large treatment plant providing high quality reuse water.

Also common to the treatment plants would be sludge treatment. Stabilization of the sludge could result in biosolids acceptable for beneficial use on agricultural land or distribution to the public.

The two main additions to the treatment train at a treatment plant to attain the high levels of effluent quality required for the different nonpotable reuse strategies are the addition of tertiary treatment and increased levels of disinfection. These two processes are discussed below.

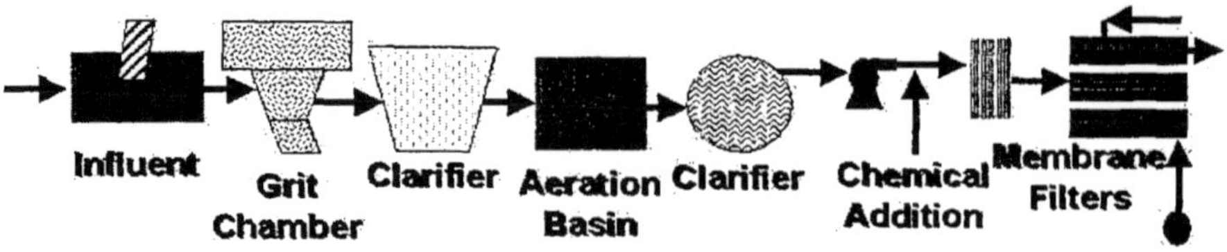

FIGURE 2 Typical large water reclamation plant configuration.

Tertiary Treatment

With the advancements in membrane technology into the wastewater area, there are now two viable media that can be used effectively to filter secondary effluent: traditional sand (or similar media) and the newer membrane materials. Filtration has been used for many years in water and wastewater applications. Some improvements have simplified the operation and reduced the head required to pass through the process. Membrane technology improvements have made their application in wastewater treatment feasible. Common activated sludge treatment followed by filtration and extensive disinfection facilities have been shown to comply with all current standards for agricultural and urban reuse categories. Both filters and membranes will also remove helminths.

A third medium is disk filtration. This process utilizes large disks of cloth media attached to rotating drums for filtration. This process is fairly new and lacks experience in large applications.

Coagulation and Flocculation

Prior to sand filtration, it may be necessary to chemically treat the wastewater to enhance the effectiveness of the filtration process. Coagulation and flocculation constitute a two-step process utilized to chemically pre-treat the wastewater prior to filtration. In reuse systems, chemical coagulation of the wastewater is typically required if the secondary effluent has turbidity greater than 5 NTU. This step is not required prior to membrane treatment.

Alum can be injected at a static mixer to coagulate the solids. The wastewater then can be treated in flocculation basins, where larger, stronger solid particles are formed. Vertical mechanical mixers slowly mix the floc and create the larger solid particles. Detention time in a flocculation basin is about 10 to 20 minutes. Wastewater with these larger solid particles then can be subjected to the filtration process.

Filtration

Filters come in many forms, but the basic concept by which they work is the same. It is a physical straining process where the solid particles pass through a bed of sand or other media. As the wastewater solids accumulate in the filter, the flow is reversed through the bed, expanding it, and allowing the solids to be backwashed out of the media. The waste backwash water is returned to the treatment plant for retreatment. The traditional sand or multimedia filter was brought to wastewater applications from the water field. The traditional high head, high rate, deep bed filter has excellent performance characteristics, producing effluent turbidity less than 1 NTU. Though the performance is compatible with the intent of eliminating all solids that could be harboring bacteria and viruses, its large size and complexity makes it mainly applicable at very large treatment facilities. Following is a discussion of other types of filtration processes.

- *High Rate-Low Head.* A filter configuration similar to that of the traditional filter is a low head, high rate, shallow bed design that has many of the same features as the traditional filter. The most significant difference in the two types of filters is that the shallow bed filter utilizes a 300 mm deep bed of sand, versus a 1,200 mm or deeper bed. This type of filter is backwashed when the head in the filter results in excessive headloss. This filter has outstanding performance but has much of the same complicated operation as the traditional filter. Also, like the traditional filter, a large quantity of waste backwash water is produced in a short period of time, requiring the capture and equalization of the return flow to the plant to prevent upsetting the main plant treatment processes.

- *Traveling Bridge (Segmented Bed).* This filter utilizes a shallow bed (300mm to 600mm) of sand or other media installed in a segmented bed arrangement. The backwash system is suspended from the bridge that moves across the bed. A pump takes filtered water from the effluent channel and directs it back into a segment of the filter bed, forcing the water back up through the filter bed, dislodging the particles removed from the wastewater. A second pump, attached to a hood that covers the segment being backwashed, draws the waste backwash water into the hood and discharges it into a trough for return to the head of the treatment plant. By continuously moving across a filter bed, this filter produces a continuous low volume of waste backwash water, eliminating the need for a separate backwash water storage basin and a basin to equalize the flow of waste backwash water returning to the main treatment plant. Significantly less operating head is needed to pass through a traveling bridge filter, allowing it to fit within many treatment plant hydraulic profiles without additional intermediate pumping into the filters.

• *Continuous Backwash.* This type of filter also requires a low operating head (450mm to 600mm). It also produces a continuous low rate of waste backwash flow to eliminate the need for flow equalization of the waste backwash return to the main treatment plant. An air lift pump continuously moves the sand media from the bottom of the filter bed to the top where the solids are separated from the sand particles and the freshly cleaned sand is returned to the filter bed. The influent is introduced at the bottom of the filter and moves countercurrent up through the downward moving sand to overflow the filter bay. The only mechanical component of this type of filter is an air compressor to produce the air needed for the airlift pumps.

Membranes

Membrane filtration also comes in many varieties. Depending on the pore opening size, the membranes remove various size particles and can produce reclaimed water of the quality required for aquifer injection or nonpotable reuse. The type most commonly applied to wastewater treatment for urban reuse has been microfiltration.

Membranes have been shown to be very effective in the removal of pathogens and viruses, to the point of approaching compliance with the disinfection standards for urban reclaimed water, but have not been shown to consistently meet the stringent coliform standards. In any case, it is common practice in reuse systems to provide a minimum of two processes to limit the potential of pass-through of viruses (known as multiple barrier treatment).

Microfiltration has a slightly slower filtration rate than filters but requires considerably greater pressure to pass through the process. Microfiltration membranes treating secondary effluent are designed to handle a loading (flux) of about 70 $l/h/m^2$. The normal operating pressure is about 1 bar, with a maximum operating pressure of 2 bars. The waste backwash water volume is comparable to that of a filter, amounting to about 5 percent of the average flow rate. Flow equalization of the waste backwash water is not required. Chemical pretreatment is also not required.

The operation of a membrane system is significantly more complicated than any of the filter systems. Auxiliary equipment required as part of the microfiltration process includes pumps, strainers, high-pressure air compressors, feed/break tanks, and a master control system. The master controller operates the system to maintain pressures, to perform intermittent air pulse cleanings, and approximately every 3 days, to perform an in-situ cleaning.

DISINFECTION

With the advances in ultraviolet light technology in wastewater disinfection, there are two viable methods that can be used effectively to disinfect tertiary effluent to the high level required for unrestricted reuse of reclaimed water. Chlorine has been used for many years in water and wastewater applications. Ultraviolet light (UV) has gained acceptance over the past few years due to its effectiveness in the reduction of coliform and viruses without creating toxic by-products.

Chlorination

Chlorine is a very effective disinfectant. When combined with the ammonia that is present in wastewater, it creates chloramines that continue to protect the quality of the reclaimed water within the distribution system.

To reach the low levels of coliform required to meet the disinfection standards, the wastewater must be in contact with the chlorine for an extended period of time. The contact time (CT) is expressed as the concentration multiplied by the detention time. To meet the disinfection standard, a CT in excess of 1,000 is recommended. For instance, for an average chlorine concentration of 10 mg/l, a chlorine contact tank must be constructed to provide 100 minutes of detention at average flow.

A chemical building with equipment sized to feed and store adequate chlorine would be required. This building would have to be specially designed to control the flow of air out of the storage room in the event of a leak. A chemical scrubber system must be installed on the exhaust air to remove any chlorine before it can be released to the atmosphere.

Ultraviolet Light

There are several different configurations of ultraviolet light (UV) systems available. They are available as low pressure-low intensity, low pressure-high intensity, and medium pressure-high intensity. The systems cover a wide range of efficient uses of the UV light produced at the most effective disinfection point of a wavelength of 254nm. The amount of maintenance required also varies with the type of system.

Of the various combinations of low or medium pressure and intensity mentioned above, the low pressure-high intensity systems provide the best utilization of the input power at the disinfecting wavelength and the lowest maintenance requirements. These systems operate at a relatively low temperature, with most light produced at the disinfecting wavelength of 254nm. This allows the system to operate efficiently, maximizing the use of the input power, thus reducing operating costs. These factors also result in lower maintenance requirements due to extended life of the lamps and other system components.

Redundant UV system capacity must be installed to allow for maintenance and reduced effectiveness with age. Spare lamps are stored on site to allow for quick replacement as lamps burn out. A small amount of chlorine must be added to the disinfected effluent to create a residual in the distribution system.

USE OF MICROFILTRATION IN THE MEMBRANE BIOREACTOR PROCESS

The increased need for reclaimed water in arid environments has resulted in the emergence of new wastewater reclamation technologies. One of the most promising and innovative technologies in water reclamation today is the membrane bioreactor (MBR) process. The membrane bioreactor combines activated sludge treatment with a membrane separation process. The reactor is operated in a similar way to conventional activated sludge (CAS), but a clarifier is not needed. Instead, a low-pressure microfilter (MF) membrane is used to perform the sludge separation. The combination of an activated sludge and membrane process produces water that has undergone secondary, tertiary and low-pressure membrane treatment using only one unit operation.

The MBR process has been shown to provide high quality effluent with high BOD removal and complete TSS reduction. Depending on the design, configuration, and need for nutrient removal, the activated sludge portion of the process can provide significant denitrification and phosphorus reduction. The MBR effluent has the added advantage of being low in turbidity, making it possible for use as feed water to reverse osmosis (RO) in industrial or indirect potable reuse systems.

Because a membrane instead of a clarifier is performing the sludge separation, the MBR can be operated at higher mixed liquor suspended solids (MLSS) concentrations and longer solids retention times (SRT). The removal of the clarifier from the treatment train eliminates such problems as sludge bulking, pin floc, and various other settling problems associated with clarifier operation. The overall footprint of an MBR system is much smaller than a CAS facility.

The membrane operation can be performed in one of two ways. An in-line MBR configuration pumps sludge from an activated sludge reactor to a pressure-driven membrane where the solids are retained and the water passes through the membrane. The membranes are systematically backwashed in order to remove solids build-up and are chemically cleaned when operating pressures become too high. Only one company currently markets the in-line configuration MBR, mainly for industrial applications.

A submerged MBR configuration has low-pressure membranes submerged in the reactor and operates under vacuum pressure. The membrane is agitated by coarse air which assists in preventing solids build up on the membrane surface. When operating pressures become high, the submerged membranes are also systematically backwashed and are chemically cleaned. There are several companies marketing the submerged MBR configuration, which makes it the most common configuration in municipal applications. Figure 3 shows a comparison of the unit processes of a conventional treatment plant and a submerged membrane MBR plant.

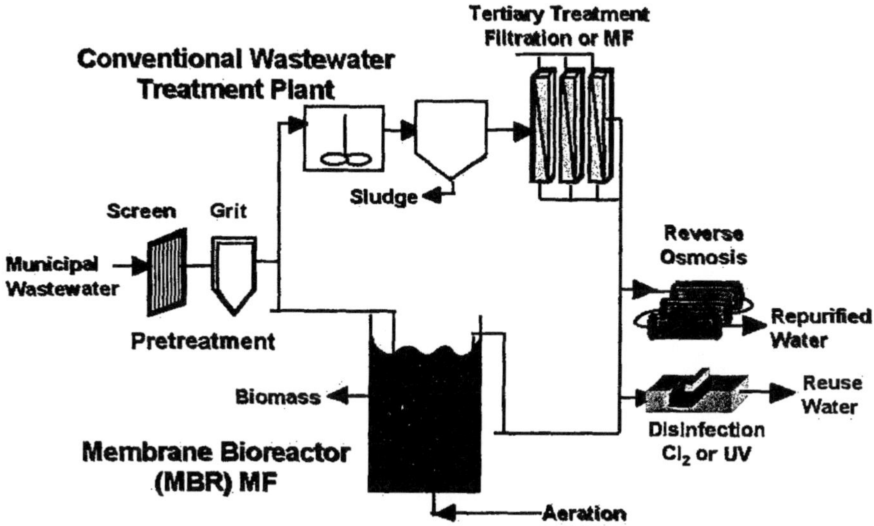

FIGURE 3 Treatment plant process comparison.

INDIRECT POTABLE REUSE DEFINITIONS, WATER QUALITY REQUIREMENTS, AND TREATMENT TECHNOLOGIES

Due to the seasonal nature of demands for nonpotable water reuse, it is likely that the supply of reuse water in a community can exceed the demand for turf irrigation and other nonpotable applications. Therefore, indirect potable reuse by aquifer recharge is a consideration for use of the excess supply of reuse water. Site-specific hydrogeologic evaluations and chemical modeling are needed to identify potential aquifer recharge processes and evaluate their suitability to the hydrogeology and the regulatory and stakeholder interests. The following discussions are focused on outlining those process options that could become part of a potable reuse strategy.

Potable Reuse

Potable reuse is subcategorized into direct potable reuse and indirect potable reuse. Direct potable reuse refers to a configuration where wastewater, treated by advanced treatment processes, is introduced directly into the potable water source for direct potable supply. Indirect potable reuse refers to a configuration where the advanced treatment waters pass through two additional stages before reaching the potable system. First, the treated waters are introduced into an environmental buffer such as a groundwater aquifer, engineered recharge gallery, surface water body, or wetland. This provides additional time and treatment by naturally occurring processes. Second, the potable water pulled from the environmental buffer also receives conventional water treatment prior to introduction back into the potable water system. Indirect potable reuse offers two benefits not found in direct potable reuse: 1) additional treatment in the environmental buffer and water supply treatment system, and 2) time to react while the advanced treatment waters pass through the environmental buffer.

Today, all active potable reuse projects in the United States employ indirect potable reuse. Direct potable reuse is currently being considered overseas and is typically an option of last resort, where pressures on available land and water resources do not allow for the additional safeguards offered by the environmental buffer. The following discussions focus on evaluating surface infiltration and indirect recharge processes in the context of the physical, regulatory, and stakeholder environment.

Aquifer Recharge

Aquifer recharge is a desirable component in the indirect potable water reuse strategy because it offers the potential for year-round storage. Further, aquifer recharge helps to maximize sustainability of a community's groundwater resource. To the degree that a community can locally replenish the aquifer, it will limit future infrastructure costs related to a declining water table and offer reliability of water supply to its residents over the long term. Aquifer recharge benefits the groundwater supply and also helps reduce depletion of rivers and other surface water resources.

It is important to recognize the public health concerns related to indirect potable reuse. The National Research Council (NRC) recently issued a report on potable reuse, entitled Issues in Potable Reuse (NRC, 1998). The report was largely supportive of indirect potable reuse, citing, as its general conclusion, "...that planned, indirect potable reuse is a viable application of reclaimed water." This report, however, does present conflicting messages on the public health risks of potable reuse, but nevertheless, contains a technical review and provides contemporary thoughts on potable reuse by the scientific community.

Water Quality Requirements of Aquifer Recharge

Water quality standards for indirect potable uses have not been formally adopted by many states. Guidelines for aquifer recharge are generally expected to maintain, at a minimum, a drinking water quality standard with additional treatment burdens for multiple barriers with stringent disinfection for pathogen removal. It is also expected that the water treated for aquifer recharge will not degrade the quality of a potable aquifer.

As an example of water quality requirements, guidelines recommended by the state of California for aquifer recharge using reclaimed water are summarized in Table 1. It should be noted that very stringent treatment and monitoring requirements are expected in the permitting process and that as more indirect reuse applications are proposed, the policies, water quality standards, and regulations are likely to change.

Indirect potable reuse applications are often permitted on a case-by-case basis dependent on the treatment technologies and monitoring requirements that are demonstrated to be protective of the public health. It is likely that water quality considerations will be the most stringent for direct injection, as this alternative places treated water directly into the aquifer and has the potential for the shortest residence times in the environmental buffer. As indicated in Table 1, reclaimed wastewater is also allowed to make up only a fraction of water actually recharged to the aquifer. Facility setback distances and residence time requirements are also likely considerations in the implementation of aquifer recharge, as these requirements provide reassurances that the environmental buffers provide additional treatment and mixing opportunities.

Treatment Technologies for Indirect Potable Reuse

Advanced water treatment beyond conventional secondary and tertiary treatment is required to remove or further reduce constituents in reclaimed water. The following constituents are targets for reductions and removal for indirect potable water uses:

- virus and pathogen removal
- nutrient (nitrogen and phosphorus) removal
- trace metals removal
- organics removal
- total dissolved solids removal.

To consider the disposal of reclaimed water into the groundwater with the possibility of that water being withdrawn in the future for domestic use, the reclaimed water must essentially meet drinking water standards. To determine the level of treatment required, pilot testing must be performed. In addition to conventional wastewater treatment processes, units to remove organics, pesticides, dissolved metals, and ultra-high level disinfection may be required. Treatment facilities in the United States are using, or considering installing, granular activated carbon (GAC), membranes providing micro- and ultra-filtration, reverse osmosis, ozonation, and ultraviolet disinfection. Treatment trains consisting of several of the above processes are required to provide multiple barriers prior to disposal of the effluent.

Figure 4 summarizes the effectiveness of advanced treatment processes and the contaminants that are removed. Figure 5 summarizes the limits and effectiveness of membrane systems to remove contaminants of concern.

TABLE 1 Sample Criteria for Aquifer Recharge with Treated Effluent

Project Category/ Level of Treatment	Maximum Percent Reclaimed Water	Depth to Groundwater		Retention Time Underground (months)	Horizontal Distance (feet)[a]
		Perc. Rate ≤ 0.20 in/min	Perc. Rate ≤ 0.33 in/min		
Surface Spreading					
Organics removal[b], oxidized[c], filtered[d], & disinfected[e]	50[d]	10	20	6	500
Oxidized, filtered, & disinfected	20	10	20	6	500
Oxidized and disinfected	20	20	50	12	1,000
Direct injection					
Organics removal, oxidized, filtered, & disinfected	50[d]	NA[f]	NA	12	2,000

[a]Horizontal Distance measured from the injection well or closest edge of the recharge basin to the nearest point of extraction.

[b]Reclaimed water used for project categories I and IV are subject to organics removal, to achieve the following product water TOC concentrations:

Reclaimed Water Contribution (%)	Maximum TOC (mg/L)	
	Category I	Category IV
0-20	20	5
21-25	16	4
26-30	12	3
31-35	10	3
36-45	8	2
46-50	6	2

[c]Oxidized wastewater is not to exceed 20 mg/L total organic carbon, 30 mg/L total suspended solids, and 30 mg/L biochemical oxygen demand.

[d]Filtered wastewater is not to exceed an average turbidity of 2 units and shall not exceed 5 turbidity units more than 5 percent of the time.

[e]For Category I, II, and IV projects, the median number of total coliform organisms in the disinfected wastewater is not to exceed 2.2 per 100 milliliters. The number of total coliform organisms is not to exceed 23 per 100 mL in more than one sample within any 30-day period.

[f]Not applicable

Contaminants Removed by Advanced Treatment Process	Suspended Solids	Pathogens	Nutrients	Metals	Total Organic Carbon	Volatile Organic Compounds	Total Dissolved Solids
Biological Nutrient Removal	●	●	●	●	●		
Coagulation-Sedimentation	●	●	●	●	●		
Granular Media Filtration	●	●					
Microfiltration and Ultrafiltration	●	●	●				
Ion Exchange			●				●
Electrodialysis			●				●
Reverse Osmosis	●	●	●	●	●	●	●
Chemical Oxidation		●			●	●	
Adsorption with Activated Carbon					●		
Air Stripping						●	
Disinfection		●					

FIGURE 4 Summary of effectiveness of advance treatment process.

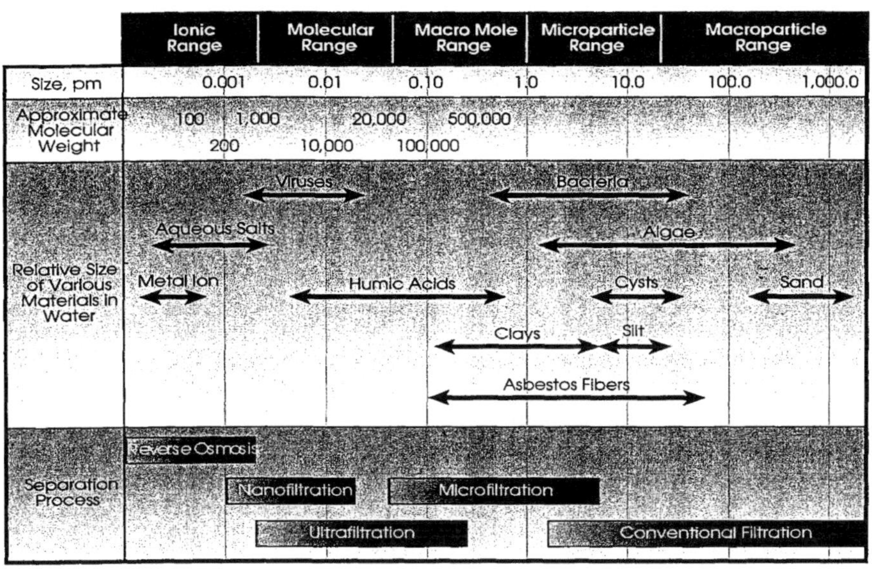

FIGURE 5 Summary of effectiveness and limitations of membrane process.

Table 2 compares conventional treatment removal mechanisms with removal mechanisms found in soil-aquifer treatment.

An advanced water treatment facility (AWTF) system is also available for treating reclaimed water for aquifer injection. Reclaimed water that has been treated with microfiltration will be delivered to the treatment site via the non-potable reclaimed water system. The primary elements of the AWTF are two-stage reverse osmosis (RO) and ozone oxidation. The RO process pressurizes the water through semipermeable membranes that reject 90 percent to 95 percent of the total dissolved solids (TDS) in the feed water. The RO membranes also provide a barrier to all microorganisms, including protozoa such as *Giardia* and *Cryptosporidium*, bacteria, and viruses. The RO membranes will also reject most of the natural organic matter (NOM) and any synthetic organic chemicals (SOCs) present in the feed water. As indicated, acid and/or a threshold inhibitor are added ahead of the RO process to lower the pH and reduce the likelihood of calcium and other ions precipitating on the membrane surfaces. Sets of cartridge filters are also included to catch any suspended material that could potentially clog the RO membrane units.

The RO permeate will receive further treatment using ozone oxidation. Ozone will provide another barrier to microorganisms. Also, ozone oxidation will provide for oxidation treatment of SOCs that could potentially pass through the RO membranes. An advanced oxidation process (AOP) can be provided by adding hydrogen peroxide or through the application of ultraviolet light following ozone addition. Sodium hypochlorite added to the water ahead of the treated water storage tank will provide a small residual in the water prior to aquifer injection. Treated water is stored for approximately eight hours in the treated water tank. The system is designed to allow the treated water to be returned to

TABLE 2 Comparisons for Soil-Aquifer Treatment

Pollutant	Conventional Treatment Removal Mechanisms	Soil-Aquifer Treatment Removal Mechanism
Virus/Pathogens	• *Coagulation* • *Flocculation* • *Filtration* • *Disinfection*	• *Precipitation* • *Adsorption* • *Filtration*
Nitrogen	• *Biological nitrification/ denitrification*	• *Aerobic and anaerobic biological degradation*
Inorganics/Metals	• *Coagulation* • *Filtration* • *Tertiary sedimentation*	• *Precipitation* • *Adsorption* • *Filtration*

the reclaimed water storage tank at the head of the AWTF if the treated water is determined not to meet the water quality requirements for aquifer injection.

A pilot study is necessary to develop design criteria prior to finalizing the AWTF process design. Additional treatment processes that may be incorporated into the AWTF include softening ahead of the RO process to remove calcium and other ions that could potentially precipitate on the membranes at higher water recovery rates. Operating the RO system at a higher recovery rate will reduce the amount of brine reject produced as well as increase the amount of water available for aquifer injection. An ion exchange could potentially be incorporated if the RO process did not adequately remove specific ions.

REFERENCE

National Research Council, Committee to Evaluate the Viability of Augmenting Potable Water Supplies with Reclaimed Water. Issues in Potable Reuse. National Academy Press, Washington, DC: 1998.

Small and Decentralized Systems for Wastewater Treatment and Reuse

Kara L. Nelson

INTRODUCTION

It is estimated that the wastewater generated by almost half of the population of the United States is treated by small or decentralized systems. Decentralized management of wastewater, which has been defined as the collection, treatment, and reuse of wastewater at or near the point of generation (Crites et al., 1998), currently serves almost one-quarter of the population. Most of this wastewater is treated at the household, although small systems that serve clusters or housing developments are becoming more common. Another quarter of the population lives in urban areas with less than 50,000 inhabitants. The wastewater generated by this population is usually collected and treated in small, centralized treatment plants.

The goal of this paper is to review the technologies that are used for the collection and treatment of wastewater from individual households and small communities, highlighting the important differences from the technologies that are used to treat larger flows. First, the significance and current status of small and decentralized treatment systems in the United States is presented. Next, implications for the reuse of wastewater at this scale are discussed. Then, the technologies and approaches for the collection of wastewater are presented. Finally, the technologies used for wastewater treatment are reviewed. Throughout the paper, recent advancements in technologies are highlighted.

SIGNIFICANCE AND STATUS IN THE UNITED STATES

Wastewater generated by the population living in rural areas is typically collected, treated, and disposed or reused at the household level using onsite facilities. In the United States, 26 million homes (23 percent of total households), businesses, and recreational facilities rely on onsite wastewater systems, which serve approximately 60 million people (USEPA, 2002) (Table 1). About one-third of new homes rely on onsite systems. The fraction of the population served by onsite systems varies widely throughout the country, with the highest fraction, 55 percent, served in Vermont, and the lowest fraction, 10 percent, served in California. It is now recognized that the fraction of the population served by onsite systems in the United States is not likely to decrease (it has not changed significantly in the past three decades), because providing centralized collection and treatment for these dispersed populations is not economically feasible.

Unfortunately, many onsite wastewater systems are failing, due to inappropriate siting, design, or maintenance (USEPA, 2002). Failing onsite systems are recognized as sources of both groundwater and surface water contamination, posing a risk to public health (due to the presence of pathogens and nitrate) and the ecological health of lakes, rivers, and estuaries (due to nutrients that cause eutrophication). The regulation of onsite systems is currently undergoing important changes, and stricter and more uniform design and performance standards are expected in the future. Many existing systems will likely be required to upgrade.

The systems that are used for the onsite treatment and disposal of wastewater in the United States typically require substantial land area. As a result, communities with a higher population density tend to have centralized collection systems that transport the wastewater to a centralized treatment plant. However, there is no specific total population, or population density, at which it is necessary to provide a sewer system. Some communities that have historically relied on onsite treatment are now installing sewer systems. For example, Chico, California, with a population of 64,000, is beginning the installation of a sewer system that will collect the wastewater for about two-thirds of its population (the rest will continue to use onsite systems), with the aim to reduce nitrate contamination of groundwater.

TABLE 1 Total U.S. Population, Population Served by Onsite Wastewater Treatment, and Population Living in Small Communities

Type of Wastewater Management	Population (millions)	Percent of Total
2001 U.S. population	285	
Onsite (individual household)		
Population served	60	23
(No. of households)	(26)	(23)
Small communities		
<10,000	29	10
<50,000	74	26
<100,000	100	35

Wastewater from most urban areas is collected and treated in centralized plants. About 10 percent of the U.S. population lives in cities with less than 10,000 inhabitants (Table 1). Another 15 percent lives in cities with between 10,000 and 50,000 inhabitants, and another 10 percent in cities with between 50,000 and 100,000 inhabitants (Table 1). The definition of what constitutes a small city (as compared to a large city) is not so important, but it is important to recognize that there are wastewater collection systems and treatment technologies available for treating small flows that are not feasible for large flows. The technologies used for small flows are highlighted in this paper.

Increasingly, decentralized wastewater management is being considered as an alternative or complement to large, centralized collection and treatment systems. Decentralized wastewater management is considered for meeting the needs of new developments within, or at the edge of, large cities (even though they already have a centralized facility). For example, the majority of new development in cities occurs at the outer edge, and as cities grow larger and larger it becomes less feasible to connect these new developments with the existing sewer network. Decentralized collection and treatment systems are becoming a more common approach for suburban housing developments.

REUSE

Small and decentralized wastewater treatment presents unique opportunities for reuse. The important characteristic that distinguishes this type of wastewater management from larger systems is that there is a much greater potential for the treated wastewater to be generated closer to the potential reuse sites. With currently available technology, the capability exists to produce wastewater at the quality that is appropriate for the specific type of reuse, ranging from irrigation of low-value crops to toilet flushing.

For onsite systems, the most common type of reuse is landscape irrigation. Given that the average person in the United States uses 170 liters/d (45 gal/d) of water outside the home, principally for irrigation, there is a large opportunity to replace the use of potable water with reclaimed wastewater for irrigation. Even if irrigation is not incorporated, it is worth recognizing that the common practice of disposing wastewater to the soil results in groundwater recharge; in some regions, such volumes may be an important part of the hydrological cycle. In-home reuse is also possible, and high quality effluent can be produced from either a part of or the entire wastewater stream.

Decentralized wastewater management, if viewed as an alternative to larger, centralized systems, presents perhaps the greatest opportunity for wastewater reclamation and reuse. For example, landscape irrigation of public areas, industrial reuse, or reuse in buildings creates a distributed demand for wastewater. If the production of reclaimed wastewater can be coordinated with the demand, facilities can be constructed close to the site of demand. This arrangement has the potential to achieve large savings in transport of both the untreated and treated wastewater. Furthermore, by treating the wastewater in smaller quantities, the necessary level of treatment can be coordinated with the reuse application. Another opportunity is for the entity reusing the wastewater to invest directly in the construction and operation of the treatment facilities. This type of arrangement is attractive to many industries or users that face difficulty finding a new or secure water source.

In small communities, often located in agricultural regions, there is a large potential for reusing wastewater for agricultural irrigation. Ironically, much of the wastewater currently generated by small communities is currently disposed of on land (spray irrigation, infiltration basins, or overland flow), but no crop is harvested. As water becomes scarcer in many regions of the country, it is likely that land disposal will be converted to planned reuse.

TECHNOLOGIES FOR WASTEWATER COLLECTION

Technologies for wastewater collection are considered in this section. First, greywater separation is discussed as an alternative management scheme for individual households. Second, alternatives to conventional sewerage are discussed that are applicable to small communities or for transporting wastewater to decentralized treatment plants.

Onsite Systems: Separate Greywater Management

The primary component of onsite wastewater collection is usually a septic tank; all of the residual water generated within the house is collected in the septic tank, which provides flow equalization as well as initial treatment. Septic tank designs, as well as alternatives for treatment and disposal are discussed in a later section. In terms of collection, however, alternative management is possible if greywater and fecal waste are managed separately. This type of management is attractive when the soil disposal of wastewater is prohibited or when there is interest in reusing the greywater, and potentially the treated fecal material, onsite. Nevertheless, some local regulations either prohibit or have ambiguous regulations for some types of greywater disposal and separate management of fecal waste.

The definition of greywater varies; typically, it is defined as the residual water produced that does not contain feces, e.g., the water from sinks, showers, dishwashers, or laundry facilities. However, local definitions may differ because of the implications for regulations. In California, for example, greywater does not include the water from toilets, kitchen sinks, dishwashers, or the laundry of diapers (Leverenz et al., 2002).

Because greywater is a low strength wastewater, with much lower concentrations of biochemical oxygen demand (BOD), nutrients, and pathogens compared to combined wastewater, it does not require as much treatment before disposal or reuse. Subsurface disposal of greywater may be possible without any treatment. However, some types of distribution systems may require particle removal to prevent clogging of small orifices. Separate collection and disposal of greywater is particularly attractive if it can be reused for landscape irrigation.

Another potential advantage of separate greywater collection is that the volumes of wastewater that require treatment or disposal may be substantially lower if fecal waste is collected by a process that does not require water. In the United States, the volume of water used to flush a toilet ranges from 1.6 gallons (6 liters) for a low-flow variety to 7 gal (26 L) for a conventional design. Up to 20 percent of a household's water may be used for toilet flushing. Thus, by not using water, the volume of wastewater is reduced, and alternatives are available for management and treatment of the fecal waste. The two main options are composting toilets and incinerating toilets. Both of these types of technologies produce a final product (compost or ash) that can be disposed as solid waste or used as a soil amendment. Many types of systems are available commercially (del Porto and Steinfeld, 1999).

Small and Decentralized Systems: Alternative Sewerage

Several alternatives to conventional gravity sewerage have been developed that may offer substantial advantages for small and decentralized communities. The most common types of alternative sewers in the United States are small diameter, pressurized, or vacuum sewers (USEPA, 1991). Of the three types of alternative sewerage used in the United States, small diameter gravity sewers (SDGS) and pressurized sewers are the most common; hundreds of these types of systems have been built, serving communities that range in size from as few as 50 households, to more than 20,000 (USEPA, 1991). A further type of alternative low cost sewerage has been developed outside of the United States, and has mostly been applied in regions where conventional sewerage is cost-prohibitive, such as Brazil, Colombia, Pakistan, India, Ghana, Zambia, and Nigeria (Mara, 1996).

All of these alternatives employ lower cost materials, typically polyvinyl chloride (PVC), because smaller diameter pipes can be used. In addition, lower slopes can be used, such that the pipes can be installed at shallower depths than conventional gravity sewers. Thus, substantial savings may be realized due to lower costs of construction (materials, excavation, and manholes). Whether alternative sewerage can be provided for lower cost than conventional gravity sewerage depends on many factors, however. For low-density developments, alternative sewerage is advantageous because the excavation and material costs are lower on a per-foot basis. In some areas, excavation to great depths for the installation of conventional sewerage is undesirable, for example if bedrock is present or if there is a high groundwater table. Finally, alternative sewerage may be lower cost if the wastewater treatment plant is located at a similar or higher elevation than the households. Any of the alternatives may provide complete sewerage for a community or may be used in combination with conventional gravity sewers, as appropriate. The main characteristics unique to each of the alternative sewer designs are briefly reviewed below.

With small diameter gravity sewers (SDGS), the wastewater from each household is treated in a septic tank before discharge to the main collector. In the septic tank, large, dense solids are removed by sedimentation. After this initial processing, the wastewater can be transported in a small diameter pipe with minimal chance for clogging, and with a lower slope because a minimum velocity does not need to be provided to prevent solids from settling out during transport. In addition, SDGS have some flexibility to follow the natural topography as long as the net downward gradient is sufficient. As with conventional gravity sewers, pump stations can be installed if the treatment plant is not located sufficiently below grade.

In a pressurized sewer system, each household has a pump and discharges wastewater to the collection system under pressure. Similar to SDGS, the wastewater also receives some processing at the household. The two options are a septic tank equipped with a septic tank effluent pump (STEP), or a grinder pump (GP), which does not require a septic tank. In addition to the advantages of the SDGS, a pressurized sewer has the additional advantage that wastewater can be transported to higher elevations, and that the pipes can follow the natural topography.

In a vacuum sewer system, the wastewater from each household is transported to an interceptor tank by gravity. Periodically, wastewater is discharged via a valve to a collector main under negative pressure, which is supplied by either one or more centralized vacuum stations. Vacuum systems have similar advantages as pressurized systems when compared to conventional gravity sewers. However, there are obvious differences in terms of system components. Some vacuum systems also have another important difference, which is the separate collection of black water (from toilets) and greywater. With this configuration, the toilets can operate under vacuum pressure, and substantial water savings can be realized because smaller water volumes are needed for flushing.

TECHNOLOGIES FOR WASTEWATER TREATMENT

For small communities and individual households, a broad range of technologies is available for treating wastewater. At one end of the spectrum are technologies that use gravity flow, have few or no moving parts, and rely on natural processes to achieve most of the treatment. These technologies tend to be lower cost, have few or no energy requirements, and require less operation and maintenance. However, they are also more dependent on climatic and environmental conditions for treatment, so the degree of treatment achieved is more variable. At the other end of the spectrum are highly mechanized technologies that use pumps to distribute the wastewater, and use mechanical equipment to provide mixing, aeration, filtration, or other augmentation. Significant advancements have occurred throughout this spectrum of technologies. In the following sections, the technologies used for onsite and small systems are reviewed as well as recent advances in technology.

Onsite Wastewater Treatment

The most common configuration for onsite wastewater treatment facilities has two components: a septic tank (ST) and a soil absorption system (SAS). Conventional ST-SAS systems in the United States are passive and operate entirely by gravity flow with no energy requirements. The purpose of the septic tank is to remove large particles by sedimentation. Two-chamber septic tanks are usually required to prevent hydraulic short circuiting. The mass of the solids that accumulates in the tank is reduced over time by anaerobic degradation. However, periodic removal of this sludge is required. Private contractors typically provide servicing of septic tanks using vacuum trucks.

One of the greatest causes of contamination from onsite systems is leaking septic tanks. It is estimated that only four to six percent of existing septic tanks in the United States are watertight. A leaky septic tank may discharge wastewater directly to the soil. Alternatively, a leaky septic tank may receive water from the soil, causing hydraulic overloading. Watertight septic tanks are now commonly available and may be manufactured from concrete, plastic, or fiberglass.

The effluent from the septic tank is discharged to the soil absorption system (SAS). The goal of the SAS is to distribute the wastewater to the soil, where it percolates through the unsaturated soil layer to the groundwater. During percolation, the wastewater undergoes further treatment by natural processes, principally adsorption to soil particles and biodegradation. A conventional SAS consists of two inch perforated pipe laid at the bottom of two foot deep, gravel-lined trenches. The discharge of wastewater throughout the SAS is usually uneven, due to the limitations of gravity flow, clogging of the orifices, and uneven settling of the SAS components.

A properly designed ST-SAS system should achieve sufficient treatment of the wastewater to prevent unacceptable contamination of the groundwater that ultimately receives the wastewater. Unfortunately, there is widespread recognition that many existing onsite wastewater treatment systems do not meet this criterion, as evidenced by the presence of fecal indicator bacteria or nitrate, or both, in groundwater wells and surface waters under the influence of groundwater. By some estimates, from 10 to 20 percent of the systems are failing, although the percentage that cause groundwater contamination may be even higher (USEPA, 2002).

In many areas of the United States, the soil type and groundwater hydrology are not amenable to a conventional SAS. In some cases, adequate disposal can be achieved by providing additional treatment of the wastewater before it is discharged to the SAS. In other cases, soil discharge is completely prohibited, and complete treatment of the wastewater must be achieved prior to surface discharge. As population pressure increases in many areas of the United States, the availability of building sites with conditions adequate for conventional ST-SAS treatment is diminishing. Thus, there is demand for treatment processes that can be used at the household level, either to augment a ST-SAS or to replace it.

Many advancements have been made that can dramatically improve the performance of existing or planned ST-SAS systems. In terms of the septic tank, the advancements include watertight tanks and ST effluent filters and pumps. The effluent filters prevent the discharge of solids that may clog the SAS or subsequent treatment processes. The effluent filter pump enables discharge to a pressurized SAS (that may be located above grade) or to another type of treatment process. Pressurized SAS can dramatically improve the distribution of wastewater to the soil, overcoming the limitations of gravity systems. Another improvement in the performance of SAS is the recognition that the upper layers of the soil have the greatest potential for treatment, as they contain a higher concentration of organic matter and higher population of soil organisms. Thus, many regions now allow the SAS to be located closer to the soil surface or even to be installed above the soil surface, provided that no direct contact with wastewater occurs.

An alternative to direct discharge of the ST effluent to the SAS is to provide additional treatment. Intermittent filters have long been used to treat ST effluent. Typical filter media are granular, such as sand or fine gravel. However, synthetic media, such as textiles sheets or open cell foam, have been demonstrated to improve performance over granular media. With all intermittent filters, improved performance has been observed at a higher dosing frequency (a smaller volume of water is applied per dose as the dosing frequency increases) for an equal surface loading rate. Wastewater may also be recirculated several times through the filter to improve performance.

An alternative to a ST-SAS is to purchase a self-contained treatment unit, often called a package plant. Over 200 types of package plants are available commercially in the United States (Leverenz et al., 2002). Most package plants employ some type of biological treatment, which may be based on aerobic, anaerobic, or anoxic conditions and use attached or suspended organisms. Other processes incorporated into package plants may include membrane filtration and disinfection by chlorine, ultraviolet light (UV), or ozone. Some package plants can produce an extremely high quality effluent and have been specifically designed for reuse.

Small and Decentralized Systems: Wastewater Treatment

Many of the same technologies that are used for treating the wastewater for large flows are also used for small communities in the United States. For example, extended aeration, oxidation ditches, and sequencing batch reactors are commonly used in small communities; all are aerobic, suspended growth biological processes and are similar to the activated sludge process. Aerobic, attached growth processes can also be used, such as the trickling filter-solids contact process. Recently, an anaerobic biological process, the upflow anaerobic sludge blanket (UASB), has been developed for treating low wastewater flows. Although this process is gaining popularity in many parts of the world, there has been little experience in the United States. A technology that has the potential for widespread application in small communities is the membrane bioreactor (MBR). This process is discussed in detail in the chapter on Large Scale Systems, and will not be reviewed here.

In contrast to the processes that are similar to those that are used for large flows, a broad group of treatment technologies that is commonly used for small communities is natural systems. In natural systems, wastewater constituents are removed or transformed by natural processes at natural rates. Thus, most natural systems for wastewater treatment require substantial land area, which often makes them infeasible for large populations. The main types of natural treatment systems can be divided into soil-based and aquatic-based processes (WEF, 2001). Soil-based natural treatment systems are as follows:

- subsurface (soil absorption system, or leachfield),
- slow rate, surface (irrigation),
- rapid infiltration (groundwater recharge), and
- overland flow.

Aquatic-based natural treatment systems are as follows:

- wastewater stabilization ponds,
- wetlands (surface, subsurface, and vertical flow), and
- floating aquatic plants (e.g., duckweed or hyacinth).

The main advantages of natural treatment systems are that they use less energy, require less operation and maintenance, and have lower construction and operation costs than more mechanized systems. The main disadvantages are that there is more variability in the effluent quality because the treatment depends on climatic factors, and that large land areas are required.

A complete description of each of these types of systems is beyond the scope of this paper. However, to illustrate some of the main characteristics of natural systems as well as the importance of these systems in the United States, wastewater stabilization ponds are highlighted in the following section. More information on each of these types of natural treatment systems can be found in WEF (2001) and Crites and Tchobanoglous (1998).

Wastewater Stabilization Ponds

Wastewater stabilization ponds (WSPs) are also called oxidation ponds or lagoons. A typical system consists of several constructed ponds operating in series; larger systems often have two or more series of ponds operating in parallel. Treatment of the wastewater occurs as constituents are removed by sedimentation or transformed by biological and chemical processes. The main biological processes are driven by bacteria and algae. Aerobic and facultative bacteria grow in the water column and consume organic matter (BOD) and nutrients. These bacteria also consume oxygen if it is present; thus, depending on the loading rate in the pond, either anaerobic or aerobic conditions will be created. Ponds that have an aerobic layer overlying an anaerobic layer are called facultative. Algae, which are present except in anaerobic ponds, also consume nutrients and play an important role in the production of oxygen that is subsequently used by the bacteria. Due to the use of CO_2, algal growth may cause the pH to rise when photosynthetic rates are high during the day, which contributes to the inactivation of pathogenic bacteria and viruses. In the bottom of the ponds, a sludge layer forms due to the sedimentation of influent suspended solids as well as the settling of algal and bacterial cells that grow in the pond. Periodic removal of the sludge may be necessary depending on the loading rates and the degree of stabilization that occurs within the sludge layer.

Depending on the configuration, pond systems are capable of achieving the equivalent of primary, secondary, or tertiary treatment. Anaerobic, facultative, or mechanically aerated ponds are used for combined primary and secondary treatment, whereas aerobic maturation ponds (with or without mixing) are used for tertiary treatment. Alternative configurations are also possible. For example, advanced, integrated wastewater pond systems (AIWPS) incorporate a deep fermentation pit into the first pond, and the configuration of the second pond is like a racetrack with mechanical mixing.

In general, the more ponds in series, the higher level of treatment. Ponds are frequently used for polishing wastewater effluent from other primary or secondary treatment processes. Pond systems are particularly effective at removing and inactivating pathogens compared to other treatment processes. Due to their long detention times, which may range from several weeks to several months, helminth eggs (such as *Ascaris*) and protozoan cysts (such as *Giardia* and *Cryptosporidium*) are efficiently removed by sedimentation. Bacteria and viruses are also removed by sedimentation if they are attached to particles. In addition, they are inactivated in the water column by a combination of sunlight-dependent mechanisms.

The main advantages of WSPs are as follows:

- provide excellent pathogen removal or inactivation,
- produce effluent well-suited to irrigation (no disinfection necessary),
- have low construction, operation, and maintenance costs,
- can be gravity fed with no moving parts, and
- require minimal technical training and skills to operate and maintain low sludge production.

The main disadvantages are as follows:

- require large land area,
- depend on climate (temperature, wind, solar irradiation) for performance and, therefore, effluent quality is highly variable,
- produce effluent that may contain high concentration of algae, and
- may discharge pathogens if the pond system is poorly designed or operated.

Pond systems continue to be used and constructed in the United States. As stated earlier, in 1983 it was estimated that there were over 7,000 pond systems in operation in the country (USEPA, 1983). More recent data have been compiled from California (California State Water Resources Control Board Database, 2000). Over 400 ponds currently exist in California, with the most popular types being aerated and facultative (oxidation) ponds (Figure 1). Most ponds are fairly

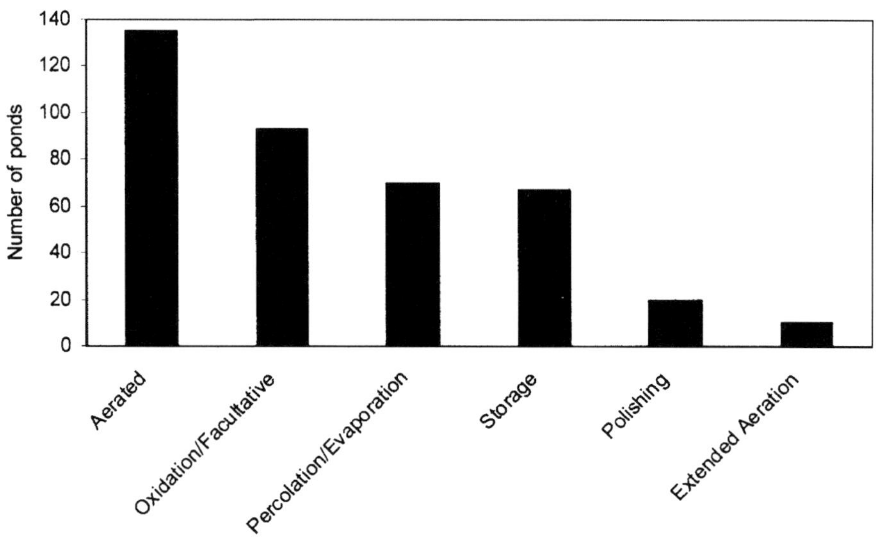

FIGURE 1 Number of ponds of each type in California.

small, with surface area less than 10 acres; however, there are approximately 80 systems with surface areas between 10 and 100 acres, and 13 systems with an area greater than 100 acres (Figure 2). The peak construction period of pond systems was in the 1970s; however, construction has continued steadily, with more than 60 systems constructed during the 1990s (Figure 3).

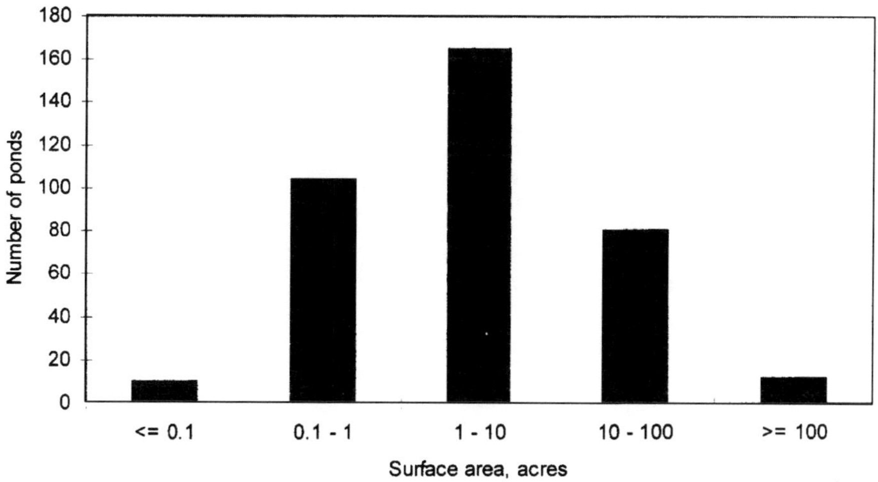

FIGURE 2 Number of ponds in California with the indicated surface area.

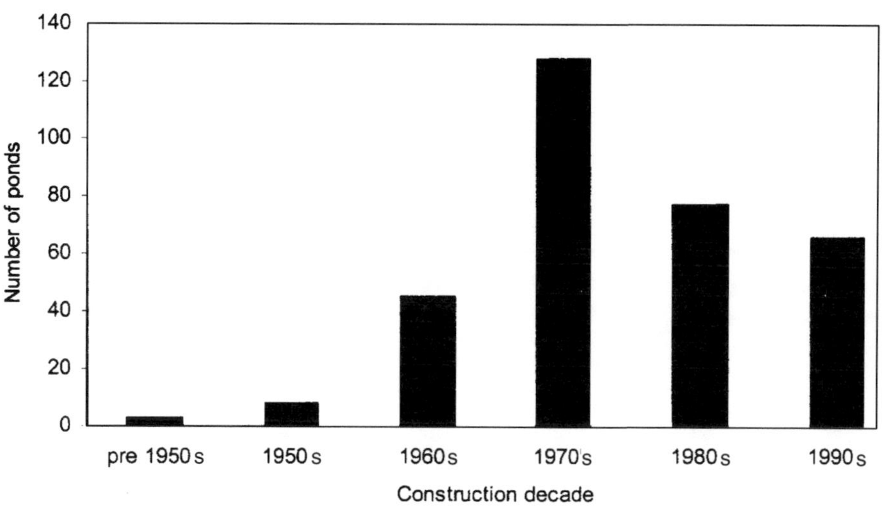

FIGURE 3 Number of ponds constructed each decade in California.

REFERENCES

California State Water Resources Control Board. Groundwater Resources Information Database. Sacramento, California 2000. *available at* HtmlResAnchor http://geotracker.swrcb.ca.gov/.

Crites, R., and G. Tchobanoglous. 1998. Small and Decentralized Wastewater Management Systems. Boston: McGraw-Hill.

del Porto, D., and C. Steinfeld. 1999. The Composting Toilet System Book. Concord, Massachusetts: Center for Ecological Pollution Prevention.

Leverenz, H., G. Tchobanoglous, and J.L. Darby. 2002. Review of technologies for the onsite treatment of wastewater in California. Report No. 02-2, Center for Environmental and Water Resources Engineering, Department of Civil and Environmental Engineering, University of California, Davis.

Mara, D., ed. 1996. Low-Cost Sewerage. Chichester, England: John Wiley & Sons.

USEPA. 1983. Design Manual. Municipal Wastewater Stabilization Ponds. EPA-62511-831015. Washington, DC.

USEPA. 1991. Alternative Wastewater Collection Systems. EPA162511- 911024. Washington, DC.

USEPA. 2002. Onsite Wastewater Treatment Systems Manual. EPA116251R- 001008, Washington, DC.

WEF. 2001. Natural Systems for Wastewater Treatment, 2nd Edition. Prepared by Task Force on Natural Systems, Sherwood C. Reed, chairman. Alexandria, VA.

Solar Desalination for Domestic Applications

Mehdi N. Bahadori

ABSTRACT

Water may be considered a more important resource than energy, given that a water crisis is life threatening. However, fresh water may be produced from sea or brackish water using energy. Large-scale sea water desalination processes are discussed, and an example project on the Persian Gulf coast is described. For small-scale fresh water production, basin type solar stills are viable options. Such stills may be designed to meet an individual family's needs or the needs of small villages scattered in the southern part of Iran. Heat transfer equations governing the operation of basin type solar stills are given, and climatological data of four cities on the Persian Gulf and Oman Sea are presented. Based on their design and operating conditions, solar stills may produce three to six liters of fresh water per square meter per day, with an estimated average production rate of 1.5 cubic meters per square meter per year ($m^3/m^2/y$) for the southern region of the country.

INTRODUCTION

Water is no longer the infinitely renewable resource that we once thought it was. In fact, water shortages threaten to make water a potentially more critical resource than energy. A water crisis, in contrast to an energy crisis, is life threatening. Unlike oil, fresh water has no viable substitute, and its depletion both in quantity and quality has even more profound economic and social effects. However, there are available techniques to produce fresh water artificially.

It is now technically and economically feasible to produce volumes of water of suitable purity through the desalination of seawater. Of course, the challenge is to produce fresh water for communities for their continuous health, development, and growth at an acceptable cost. To meet the challenge, large desalination systems, including dual-purpose power and desalination plants, have been built to reduce the cost of production of electricity and water. Thermal energy extracted or exhausted from power plants is used effectively in the desalination process. It is estimated that there are over 25,000 megawatts (MW) of power combined with desalination plants used in the cogeneration concept. However, not all water demands are coupled with the need for additional electric power.

A worldwide inventory shows that by the end of 1995 there were over 11,000 desalting units with total capacity of 20 million cubic meters per day. Desalination is already used in 120 countries around the world. The exponential growth of desalination can be illustrated by the fact that in 1971 total worldwide capacity was only 1.5 million cubic meters per day. In 1976 the total was 4 million cubic meters per day and in 1995 it was 20 million cubic meters per day. In the last 10 years, worldwide capacity grew from 12 to about 22 million cubic meters per day.

The Middle Eastern countries are the biggest users of desalination technology, with about 50 percent of the world's capacity installed in the area. The dominant plant type is Multi Stage Flash (MSF), which accounts for 86.7 percent of the desalting capacity, while the Reverse Osmosis type accounts for only 10.7 percent. In the state of Hormozgan, in the southern part of Iran, about 45 percent of the fresh water produced from the sea for the cities and islands is through MSF, 20 percent through Multi Effect Distillation (MED), 31 percent through Vapor Compression (VC), and 4 percent through Reverse Osmosis (RO). Worldwide, 48.1 percent of the total installed or contracted capacity is based on the MSF principle, reflecting a continuing decline from the proportion reached in 1991 (51.5 percent). In comparison, the Reverse Osmosis process increased its share from 32.7 percent to 35.9 percent in the same period.

LARGE-SCALE DESALINATION PROCESSES

Desalination can be classified into phase-change and single-phase processes. The most commonly used phase-change processes are Multi Stage Flash (MSF), Multi Effect Distillation (MED), Vapor Compression (VC), and Solar Distillation. Highly developed single-phase processes are Reverse Osmosis (RO) and Electrodialysis (ED), which use membranes to separate impurities from water (Assimacopoulos, 2001).

Solar energy may be employed to produce fresh water from the sea. This may be accomplished in a large system or in a simple basin type desalination unit. For a large quantity of fresh water production, a unit was constructed in the city of Abu Dhabi on the Persian Gulf coast using solar energy (El-Nashar, 2001). The plant consists of three subsystems: the solar collector field, the heat accumulator, and the sea water evaporator. It is designed for an expected yearly average fresh water production of 85 cubic meters per day, using sea water with a salinity of 55,000 ppm or 5.5 percent (El-Nashar, 2001).

The solar energy collecting subsystem converts solar energy into thermal energy when solar radiation is available during the day, using a bank of solar collectors. The thermal energy is stored in the heat accumulator subsystem for heating the evaporator with minimum fluctuation in the supply temperature.

The basic unit in the bank of solar collectors is an evacuated tube panel that employs selective coating absorber plates enclosed in glass tubes maintained under high vacuum of 10 mm Hg. Ten glass tubes with their absorber plates are incorporated in each panel. Along the center line of each glass tube is a single copper tube, which is attached to the middle of the absorber plate. Heat collecting water flows through this center pipe and absorbs the solar energy collected.

Each panel has an absorber area of 1.75 square meters, and the selective coating on the absorber plates has an absorptivity of 0.91 and an emissivity of 0.12. The collector bank consists of 1,064 panels making up a total collector absorber area of 1,862 square meters (El-Nashar, 2001).

In order to make solar desalination more efficient, a horizontal tube, a thin film multiple effect evaporator is used with a rated capacity of 120 cubic meters per day. Preheated feedwater is sprayed on the top of the first effect tube bundle and descends down the evaporator stack, flowing as a thin film over each following effect tube. The feedwater flashes and thereby is cooled by several degrees as it passes from one effect tube to the next. It is rejected at the bottom of the last effect tube as cool concentrated brine, which is discharged to the sea. In the top effect tube, heating water from the accumulator is used to partially evaporate the thin seawater film on the outside of the tubes (El-Nashar, 2001).

BASIN TYPE SOLAR STILLS

The process discussed above, and other desalination processes mentioned earlier, are primarily for large-scale fresh water production using seawater. There are many small villages or communities that use small quantities of fresh water. For small-scale fresh water production, using sea or brackish water, simple basin type solar desalination seems to be a viable option.

To see the need for small-scale fresh water production, we may consider the state of Hormozgan again. About 60 percent of the population of this state lives in over 2,000 villages, each village housing about 50 families. These families draw their water needs from wells, rivers, cisterns, or other reservoirs. The water drawn from such sources is often slightly salty and many times not hygienic. To meet the fresh water needs of each family, or a village of about 50 families, simple solar desalination or distillation systems may be employed.

Basin type solar distillation is the simplest desalination process and is based on the greenhouse effect. Glass and other transparent materials have the property of transmitting incident short-wave solar radiation but do not transmit infrared radiation. Incident short-wave solar radiation passes though the glass into the still where it is trapped and evaporates the water, which is then condensed on the glass surface and is collected as distillate. The equipment is simple to construct and operate. However, a large area of land is required.

The first known application of solar distillation was in 1872, when a still at Las Salinas on the northern deserts of Chile started its three decades of operation to provide drinking water for animals used in nitrate mining (Duffie and Beckman, 1981). The still utilized a shallow black basin to hold the salt water and absorb solar radiation; water vaporized from the brine, condensed on the underside of a sloped transparent cover, ran into troughs, and was collected in tanks at the end of the still. Most stills built and studied since then have been based on the same concepts, though many variations in geometry, materials, methods of construction, and operation have been employed (Duffie and Beckman, 1981).

A basin type still is shown in section in Figure 1. A solar desalination plant may have many bays side by side, each of the type shown. The covers are usually glass; they may also be air-supported plastic films. The basin may be on the order of 10 to 20 mm deep (referred to as shallow basins), or they may be 100 mm or more deep (referred to as deep basins). The widths are on the order of 1 to 2 m, with length widely variable up to 50 to 100 m (Duffie and Beckman, 1981).

In operation, solar radiation is transmitted through the cover and absorbed by the salt water and basin. The solution is heated, water evaporates, and vapor rises to the cover by convection, where it is condensed on the under side of the cover. Condensate flows by gravity into the collection troughs at the lower edges of the cover; the covers must be at sufficient slope so that the surface tension of the water will cause it to flow into the troughs without dripping back into the basin. The trough is constructed with enough pitch along its length so that the condensate will flow to the lower end of the still, where it drains into a product

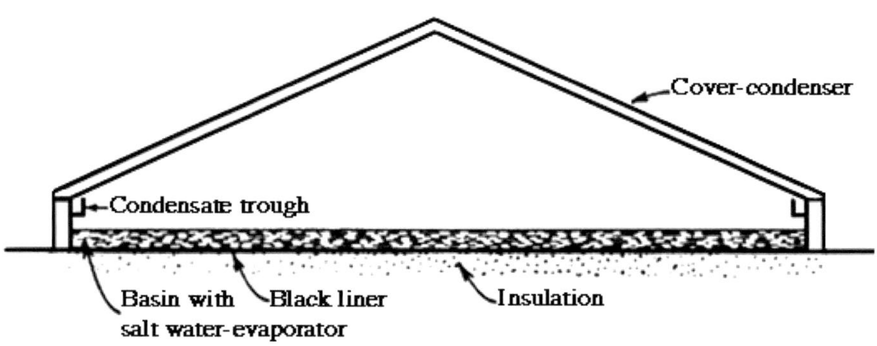

FIGURE 1 Schematic cross section of a basin type solar still.

collection system. Operation of a still may be continuous or batch. If sea water (approximately 3.5 percent salt) is used as feed, the concentration is usually allowed to double before the brine is removed, so about half of the water in the feed is distilled off.

The glass cover may be sloped between 15 and 30 degrees. However, with special treatment of the glazing by hydrofluoric acid or sodium silicate, the surface can be made more wettable (Bahadori and Edlin, 1973). With such treatments of the glass cover, it is possible to reduce the slope of the cover down to 3 degrees and increase the fresh water production of the solar still between 50 and 70 percent (Bahadori and Edlin, 1973). The fresh water production rates of simple basin type solar stills are between 3 and 8 liters per square meter per day, depending on the design and the operating climatic conditions, particularly the solar radiation intensity.

GOVERNING EQUATIONS IN BASIN TYPE SOLAR STILLS

Figure 2 shows the major energy flows in a still while it is operating. The objective of still design is to maximize Q_{evap}, the transport of absorbed solar radiation to the cover-condenser by water vapor, as this is directly proportional to the still productivity. All other energy transfer from basin to surroundings should be suppressed, as far as is possible. Most energy flows can be evaluated from basic principles, but terms such as leakage and edge losses are difficult to quantify and may be lumped together and determined experimentally for a particular still (Duffie and Beckman, 1981).

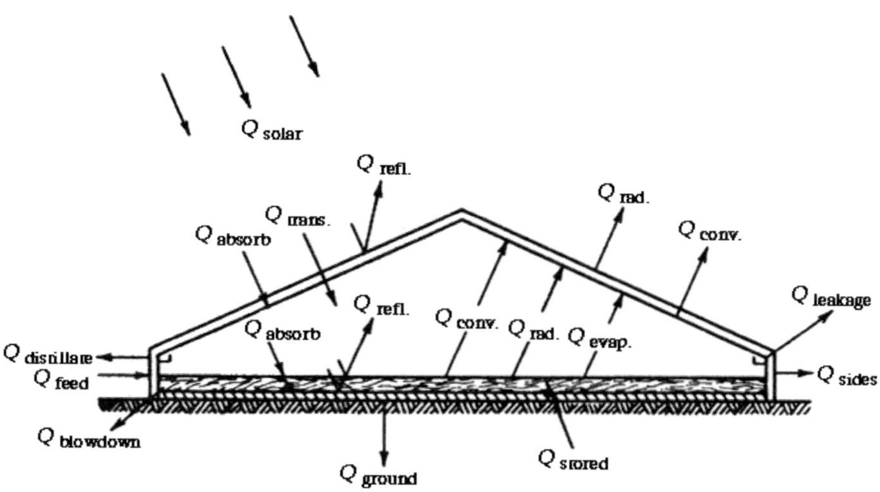

FIGURE 2 The major energy transport mechanisms in a basin type still.

Energy transfer from basin to cover occurs by evaporation and condensation, in addition to convection and radiation. The losses from the back of the still are to the ground. The depth of the water in the still is usually such that its capacitance must be taken into account.

A thermal network is shown in Figure 3, where the resistances correspond to the energy flows in Figure 2. Terms for leakage, edge losses, entering feedwater, and leaving brine or product are not shown (Duffie and Beckman, 1981).

The convection heat transfer coefficient in a still is (Duffie and Beckman, 1981)

$$h_c = 0.884\left[T_b - T_g + \left(\frac{p_{wb} - p_{wg}}{2016 - p_{wb}}\right)T_b\right]^{1/3} \quad (1)$$

where h_c is the heat transfer coefficient, T_b is the brine water temperature in degrees Kelvin, T_g is the glazing temperature, p_{wb} is the saturation pressure of water at the brine temperature, and p_{wg} the saturation pressure of water at the glazing temperature, both in mm Hg.

The heat transfer between the basin and cover is

$$q_{c,b-g} = h_c\left(T_b - T_g\right). \quad (2)$$

By analogy between heat and mass transfer, the mass transfer rate can be written as

$$\dot{m}_d = 9.15 \times 10^{-7} h_c\left(p_{wb} - p_{wg}\right) \quad (3)$$

where $\dot{m}_d$ is the mass transfer rate in kg/m^2s.

The heat transfer by evaporation and condensation is

$$q_e = 9.15 \times 10^{-7} h_c\left(p_{wb} - p_{wg}\right)h_{fg} \quad (4)$$

where h_{fg} is the latent heat of water, in J/ kg. Equation 3 shows that fresh water production is directly proportional with ($p_{wb} - p_{wg}$). While it is highly desirable to have a very high brine temperature (T_b), it is also necessary to have as low a glazing temperature (T_g) as possible.

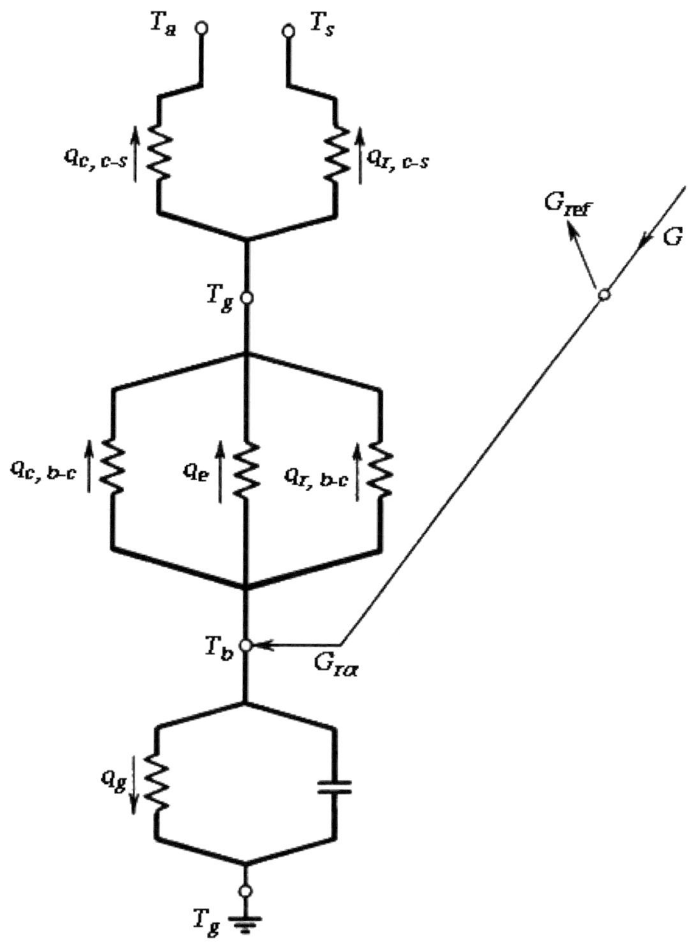

FIGURE 3 Basic thermal network for a basin type still.

If the still has insulation under the basin, heat loss to the ground can be written

$$q_k = U_g\left(T_b - T_a\right) \tag{5}$$

where U_g is an overall loss coefficient to ground, assuming the ground to be at a temperature equal to the ambient (T_a). This term will be small in a well-designed still.

The efficiency of a still is defined as the ratio of the heat transfer in the still by evaporation and condensation to the radiation (G) incident on the still.

$$\eta_i = \frac{q_e}{G}. \tag{6}$$

This is usually integrated over some extended period (e.g., day or month) to indicate long-term performance. If there is any loss of product water back into the basin (by dripping from the cover or leakage from collecting troughs), less product will be available than is indicated by this equation. Efficiency from experimental measurements is

$$\eta_i = \frac{\dot{m}_p h_{fg}}{G} \tag{7}$$

where $\dot{m}_p$ is the rate at which distillate is produced from the still (which may be less than $\dot{m}_d$), and h_{fg} is the latent heat of vaporization.

The objective of still design is to maximize q_e (i.e., $\dot{m}_d$), which is proportional to the vapor pressure difference between basin and cover. Thus it is desirable to have the basin temperature as high as possible, which will increase the ratio of heat transfer by evaporation and condensation to that by convection and radiation. Shallow basins with small heat capacity will heat up more rapidly than will deep basins, and operate at higher mean temperatures.

Many practical considerations govern solar still design and operation. Shallow basins require precise leveling of large areas, which is costly. Crystals of salt build up on dry spots in basins, leading to reduced overall absorbance and reduced effective basin area. Leakage can cause problems in three ways: distillate can leak back into the basin; hot salt water can leak out of the basin; and humid air from inside the still can leak out through openings in the cover. Occasional flushing of still basins has been found to be necessary to remove accumulations of salt and organisms such as algae that grow in the brines. Growths can be controlled by additions of algaecides (Duffie and Beckman, 1981).

A wide variety of experimental basin type stills has been built and studied. Two design trends have evolved. Large area deep basin stills, which can be built by standard construction techniques, are durable and are relatively inexpensive. Modular shallow basin stills have lower thermal capacitance, produce somewhat more water, but may be more expensive to construct (Duffie and Beckman, 1981; Bahadori and Edlin, 1973).

The base of the basin should have the property of absorbing solar radiation as much as possible. Black paint may be employed on this surface. A problem with such paints occurs during the period when the stills are left in the sun with no seawater in them. The temperature of the paint may increase greatly, thus destroying the paint. To remedy this problem, a thin layer of black pebbles may be employed at the bottom of the basin. This increases the water evaporation rate from the basin to some extent.

RECENT EXPERIMENTAL RESULTS

Small simple basin type solar stills, built of stainless steel and insulated on the sides and bottom, were tested in Tehran and Bandar Abbas. These stills had one and two layers of glazings on top. Their fresh water production rates varied between 4 and 5.9 liters per square meter per day (liter/m^2/day), during August and September 2002, under clear sky conditions (Haghbin, 2002). The still with double glazing produced more fresh water than the one with single glazing. Figure 4 shows the experimental setup in Tehran (Haghbin, 2002).

FIGURE 4 The experiment setup of solar stills built of stainless steel.

Similar experiments were carried out in Tehran, employing small stills built of plexiglass. No insulations were employed on the sides of the stills, and the condensate produced on the side walls of the stills was also collected. The fresh water production of these stills varied between 3.5 and 4.5 liter/m^2/day during the month of August 2002, under clear sky conditions (Ahmadi, 2002). Different materials, including black paint and small pebbles, were employed at the bottom of the stills. It was found that the still with small black pebbles performed best. Figure 5 shows the experimental setup, and Figure 6 the efficiency of the still, using black pebbles at the base of the still (Ahmadi, 2002). It is estimated that one can produce about 1.5 m^3/m^2/y of fresh water in the southern region of Iran by simple basin type solar stills.

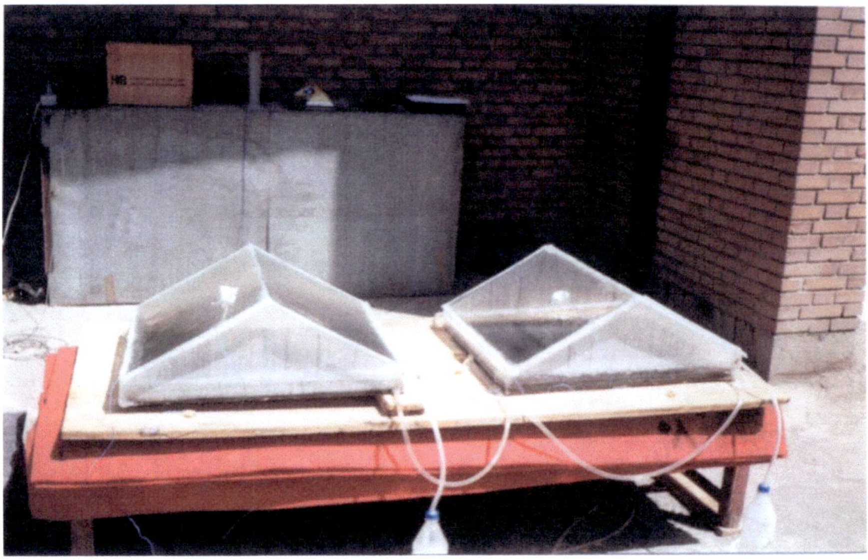

FIGURE 5 The experiment setup of solar stills built of Plexiglass.

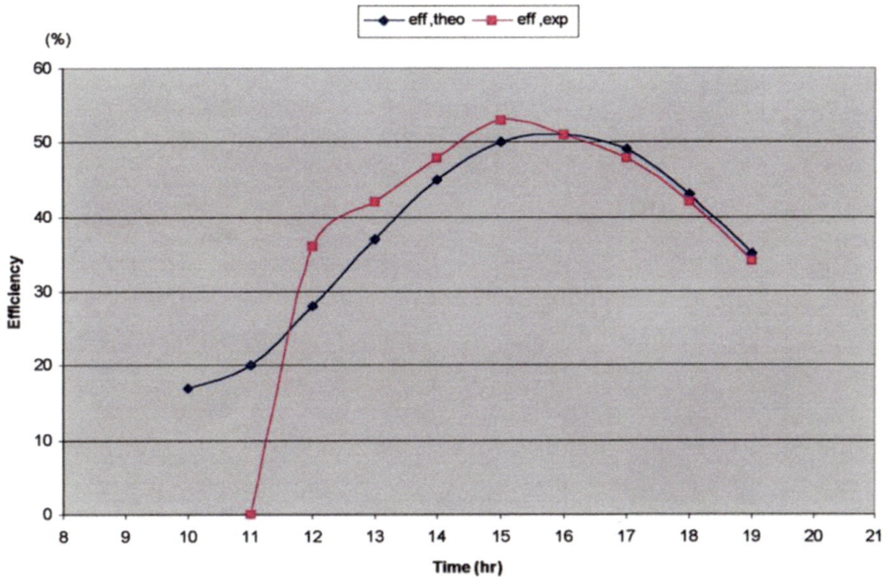

FIGURE 6 Experimental and theoretical values of efficiencies for the Plexiglass solar still, using a layer of black pebbles at the bottom of the still. Values for Tehran during August 2002.

REFERENCES

Ahmadi, N. 2002. Experimental-analytical evaluation of the performance of solar desalination designs. B.Sc. Project, Sharif University of Technology.

Assimacopoulos, D. 2001. Water water everywhere, desalination powered by renewable energy sources. REFOCUS (July/ August), Pp. 38-43.

Bahadori, M.N. and Edlin, F.E. 1973. Improvement of solar stills by the surface treatment of glass. Solar Energy 14:339-352.

Duffie, J.D. and W.A. Beckman. 1981. Solar Engineering of Thermal Processes. New York: John Wiley and Sons.

El-Nashar, A.M. 2001. Water from the sun—Case study: the Abu Dhabi solar desalination plant. REFOCUS (March), Pp. 26-29.

Haghbin, M. 2002. Solar Desalination Project. Final report submitted to Water Resources Management of Iran, Ministry of Power, Tehran.

AGRICULTURAL WATER USE AND DROUGHT MANAGEMENT

Optimal Irrigation: Considerations for Semiarid Regions

John Letey

Irrigation is required for high crop productivity when precipitation is inadequate to meet the crop evapotranspiration (ET) demand. The soil provides a storage capacity for water from which the crop withdraws water. The overall quantitative storage capacity depends on both the type of soil and the rooting depth of the crop plants. Soils whose texture and structure result in large pore sizes have lower storage capacity than soils with smaller pore sizes. More deeply rooted crops have a higher storage capacity as compared to crops whose roots are shallower. In principle, the amount of irrigation applied should provide for recharge of the soil storage capacity between irrigation events. The time to irrigate is before the soil has become too dry to allow good crop production. The combination of a soil and crop rooting system that have a low storage capacity dictates the use of smaller-quantity, more frequent irrigations, whereas larger storage capacity of soil and crop rooting systems allow less frequent though higher quantity irrigations.

The salinity of the irrigation water affects irrigation management. Plants transpire pure water, causing the soil solution to become concentrated with salts as transpiration proceeds. Because of this effect, irrigation amounts must not only recharge the storage capacity, but additional water may be necessary to leach excessive salts from the root zone.

Irrigation uniformity adds complexity to irrigation management. A uniform irrigation is one in which the same amount of water infiltrates the soil at all points in the field. Most frequently, the amount of water that infiltrates into soil is variable for different parts of the field. Nonuniform irrigation creates a dilemma. If irrigation is programmed to restore the storage capacity in the parts of the field that receive the most water, the other parts of the field will be underirrigated, causing yield reduction in those underirrigated areas. Conversely, if irrigation is programmed to recharge the storage capacity for those zones that have the lowest infiltration rate, the other parts of the field will be excessively irrigated leading to unrecoverable water lost to deep percolation. Uniformity of irrigation, therefore, is one of the most critical factors affecting irrigation management.

IRRIGATION SYSTEMS

Irrigation systems can be broadly categorized as being pressurized or nonpressurized. Pressurized systems are those in which water is delivered through a pipe, under pressure, and discharged by one of a variety of different outlet designs, including sprinkler heads and drip emitters. Nonpressurized systems are those in which water is delivered in an open channel and allowed to flow across the field. Various configurations for flow across the field, such as furrows or basin borders, are possible. The advantage of pressurized systems is that the quantity of water applied to the field can be precisely controlled by valves. Furthermore, if the system is properly designed, all the water infiltrates into the soil with no runoff from the field.

Although the amount and timing of water delivery to a nonpressurized system can be controlled, the amount of water that actually infiltrates the soil depends on the soil properties, among other conditions. Although the irrigator has some design control features for nonpressurized systems, such as length of furrow, rate of water discharge, and time (duration) of water application, the quantitative control is limited. Runoff from the field is usually an unavoidable condition for nonpressurized systems. Water must be maintained on the soil surface at the lower end of the field to allow adequate infiltration. During that period of time water is flowing off of the field.

Properly designed and maintained pressurized systems can deliver water very uniformly across the field. However, if water is emitted into the air, such as through sprinkler systems, the wind currents can greatly affect the distribution of water. Therefore, although sprinkler systems can theoretically be designed to be very uniform, wind can cause a very nonuniform distribution. Pressurized systems that emit water without spraying into the air, such as drip systems, allow for very uniform irrigation.

Nonpressurized irrigation systems have two sources of nonuniformity. First, water is on the soil for a longer period of time at the top end of the field as compared to the lower end of the field. This provides the opportunity for more water to infiltrate at the top end as compared to the lower end of the field. The term for this kind of nonuniformity is opportunity time nonuniformity. Variability of soil infiltration rate across the field due to textural or structural changes also leads to nonuniform infiltration across the field. The total nonuniformity, therefore, is a combination of the opportunity time and the soil variability.

Accurate measurement of irrigation uniformity is important in developing the optimal management scheme for a given irrigation system. Unfortunately, measurement of uniformity is complex. Uniformity of sprinkler systems is measured by distributing containers throughout a collection area and measuring the amount of water collected in each container. The data are then statistically analyzed for variability. The numerical result depends on the size of the container. Using larger containers will result in a higher uniformity value than using smaller containers under the same conditions. Thus, the value is already recognized as being somewhat subjective based on the measuring technique. Even a drip system is very nonuniform if the measurement is made on a very small scale. The amount of soil water is high at the location adjacent to where water is released from the emitter, and the amount of soil water between the emitters is very low. Nevertheless, depending on the type of plant, the plant root system can integrate differences in soil water content in different parts of the root zone and even things out. This factor raises an additional point concerning uniformity. The plant root system can accommodate uneven water distribution and can extract more water where the soil water content is high, as long as the scale of the uneven soil water distribution is proportional to the scale of the plant root system. A tree with a large root system can accommodate considerable nonuniformity of water application under the canopy. A shallow-rooted vegetable crop would be more highly impacted by the same distribution.

The nonuniformity for surface systems is determined by measuring the rate of advance of water down the furrow, and then inserting these numbers into equations developed to compute the nonuniformity associated with opportunity time. These measurements do not include the nonuniformity associated with soil variability, which can be considerable. Therefore, the numerical values for furrow systems are overestimates of the true uniformity.

As long as the measurement procedure for a given irrigation system is consistently used, the comparative uniformity of different fields for that particular system can be determined. In other words, the uniformity from one sprinkler system can be compared to a different sprinkler system. However, it is not appropriate to compare a uniformity coefficient that has been measured for a furrow system to a sprinkler system.

With all of the factors considered, a drip system has better uniformity than a sprinkler system, which is better than a nonpressurized system. However, the costs for these systems are in the reverse order. The improved performance from a drip system compared to a pressurized system may not justify the costs for the drip system. Furthermore, pressurized systems require an energy supply that may not always be present in the field.

PLANT RESPONSE TO WATER

A major portion of the sun's energy striking leaf surfaces is dissipated by evaporation from plant surfaces (transpiration). As long as the plant's roots can extract water from the soil and transfer it to the leaf surfaces at a rate equal to the transpiration rate, the plant can grow at its potentially highest rate. If the rate of transfer of water from the soil through plant tissue to the leaf surface is less than the plant's loss of water by transpiration, the plant responds by closing stomata to reduce the water loss. Carbon dioxide (CO_2) used for plant photosynthesis enters the plant through these same stomata. Therefore, closure of the stomata to reduce water loss also reduces CO_2 intake and, therefore, reduces the rate of photosynthesis. Thus, the plant has a dual mechanism for protecting itself under a limited water supply. It reduces transpiration, thus attempting to maintain turgidity. At the same time, it reduces photosynthesis, which would otherwise increase the plant surface area through growth, leading to greater interception of energy and more transpiration in a continuing cycle.

Numerous studies have reported that total dry matter production in plants is linearly related to evaporation (ET). This observed result is consistent with the mechanisms that the plant uses to protect itself against inadequate water. However, the marketable product of some crops is not linearly related to total dry matter production. Therefore, the relationship between production of the marketable product and ET must be established independently for individual crop species.

WATER USE EFFICIENCY AND WATER CONSERVATION: DEFINITIONS

Water use efficiency and water conservation are commonly used terms in irrigated agriculture. However, confusion can arise because these terms can have multiple definitions. Water use efficiency is calculated as the ratio of two terms. Yet, different measurements may be used for the two terms in the ratio, resulting in different numbers, and yet all are referred to as water use efficiency. For example, water use efficiency can be defined as the ratio of beneficial water use to applied water. However, beneficial use sometimes is represented by ET, and at other times it is represented by ET plus the amount of water required for leaching salts from the root zone. Some individuals include all of the water delivered to the field as applied water, whereas others might only consider the infiltrated water that would be available for crop use. Obviously, different numbers result for any combination of these terms. Furthermore, the computations can be made on different land area sizes. For example, the ratio can be calculated for a field, the total farm, or the total basin. Different numbers result depending on which is selected. Possibly the biggest problem, however, is the common belief that a higher efficiency number is always better than a lower number. As will be discussed later in this paper, this is not usually the case.

Water conservation likewise is subject to different definitions. One definition is to use less water. This can be accomplished by various means, each with a specific consequence. A farmer can use less water by not growing a crop. Or, a farmer can grow a crop and apply a small amount of water resulting in very low crop production. Farmers can also grow a crop for high production and either eliminate runoff or capture runoff and use it as part of the irrigation supply.

The main point is that in using these terms, the definition must be clearly specified and the consequences of the action properly evaluated. A further complication is that all of the definitions are based on water quantity without reference to water quality, and water quality cannot be ignored in a management scheme.

CROP-WATER PRODUCTION FUNCTIONS

Agricultural production is a business operation, and irrigation management must be evaluated in context of the business. The goal of any business is to maximize profits. Maximizing profits can include sustaining the business through a period when profits are not possible and generating growth in anticipation of future profits. Therefore, one definition of optimal irrigation management is that management which maximizes profits.

Water is an input to the production system. The functional relationship between crop yield that is marketed and the amount of water applied must be known for the economic analysis. Only water that infiltrates into the soil has an opportunity to contribute to crop production. The water running off the field cannot contribute to crop production, so the crop-water production function can only be based on the amount of infiltrated water. The runoff water has economic implications that must be accounted for separately.

The uniformity of irrigation significantly affects the crop-water production function. The relationship between cotton lint yield and the amount of infiltrated water for various irrigation uniformities is presented in Figure 1 for climatic conditions of the San Joaquin Valley of California. The uniformity is characterized by the Christiansen uniformity coefficient (CUC), where a value of 100 represents perfectly uniform irrigation and decreasing values of CUC represent increasing nonuniformity. Also depicted in Figure 1 is the amount of water that would percolate below the root zone for the various irrigation uniformities and infiltrated water amounts.

Considering the uniform irrigation first, crop yields increase with increasing amount of infiltrated water until a maximum yield is achieved and additional applied water does not contribute to more production. When the irrigation is nonuniform, more average water must be infiltrated for the field to get the highest yields. For a given amount of infiltrated water, the yield decreases as the irrigation uniformity decreases.

Under uniform irrigation, no deep percolation occurs until enough water has been applied to reach maximum yield. In other words all of the water applied is used by the crop. After maximum yield has been achieved, any additional applied water goes directly to deep percolation. In contrast, for nonuniform irrigation

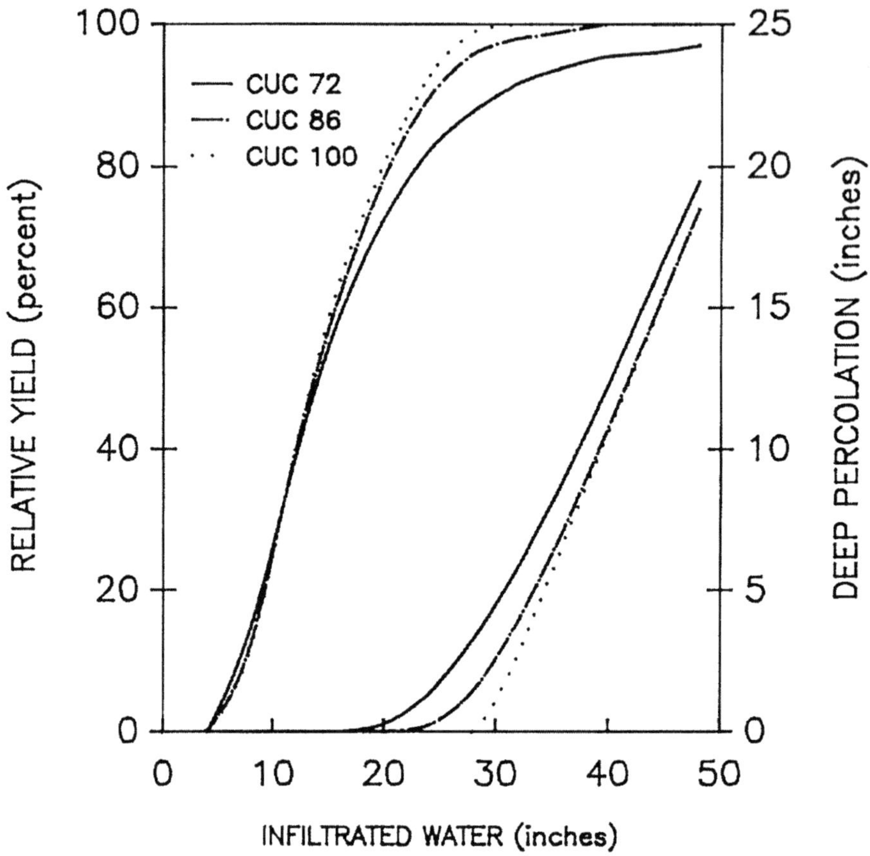

FIGURE 1 The relationships between relative yield of cotton lint and the amount of deep percolation to the amount of infiltrated water for different irrigation uniformities as depicted by Christiansen uniformity coefficient (CUC) values.

some deep percolation occurs before a maximum yield can be achieved. This is a consequence of some parts of the field having more water than necessary for maximum crop yield and other parts of the field having less water than required. In all cases, increasing water application increases the amount of deep percolation. For nonuniform irrigation there is a tradeoff between irrigating for high crop yield and low deep water percolation.

The salinity of the irrigation water is another factor that affects the crop-water production function. The salts in the irrigation water become concentrated through evapotranspiration and, therefore, some water to leach the excess salts from the root zone must be applied. The amount of water to be applied, however, depends on the salinity level of the irrigation water and the crop sensitivity to salinity.

Letey et al. (1985) developed a model to compute relative crop yield in relationship to the seasonal applied water for crops with different tolerance to salinity. The crop-water production functions are presented in Figure 2 for corn, which is a salt sensitive crop, and for cotton lint, which is a salt tolerant crop. The applied water (AW) assumes that all of the water infiltrates, and E_p is the pan evaporation. The ratio AW/ E_p may be used to facilitate comparisons between different climatic zones with different E_p values.

Note as the salinity of the irrigation water increases, more water must be applied to get the same yield. Irrigation water salinity may get to a level where maximum yield cannot be achieved regardless of the amount of water applied. Larger differences in yield for a given amount of applied water, or larger differences in applied water for a given yield, occur for the salt sensitive corn than for the salt tolerant cotton. Indeed for cotton, irrigation water salinities up to 4 deciSiemens per meter (dS/m) require relatively small amounts of additional water to achieve the maximum yield.

All of the curves depicted in Figure 2 assume that the irrigation is uniform. Nonuniform irrigation would modify the results in a manner as depicted in Figure 1. In other words, increasing nonuniformity would cause each of the curves on Figure 2 to be lowered for a given water application.

ECONOMIC IRRIGATION EFFICIENCY

Economic irrigation efficiency will be defined at the farm level as the irrigation management that maximizes profit. In a broader context, economic irrigation efficiency could be defined as irrigation management that maximizes net social benefits. The difference between the two definitions is the result of externalities. An externality arises when some of the costs or benefits of irrigation agriculture accrue to society as a whole and the costs (as reflected in market prices) are not borne by the farmers or the consumers of their products. Externalities can be positive or negative. An example of a positive externality occurs when water purchased by a farmer runs off his farm and serves some beneficial societal use. However, if the water is polluted, it can impose a cost to society and create a negative externality.

The crop-water production function as depicted in Figures 1 and 2 can be converted to benefit curves by multiplying the yield and the market price for the crop. Such curves are depicted in Figure 3 for a hypothetical case representing two irrigation uniformities. The total benefit (TB) in \$ ha^{-1} for a given infiltrated water (IW) is higher for the more uniform (TB1) than the less uniform irrigation (TB2). The total cost of water (TC) is also depicted in Figure 3 for two water prices, where the price of water for case 1 (TC1) is greater than for case 2 (TC2).

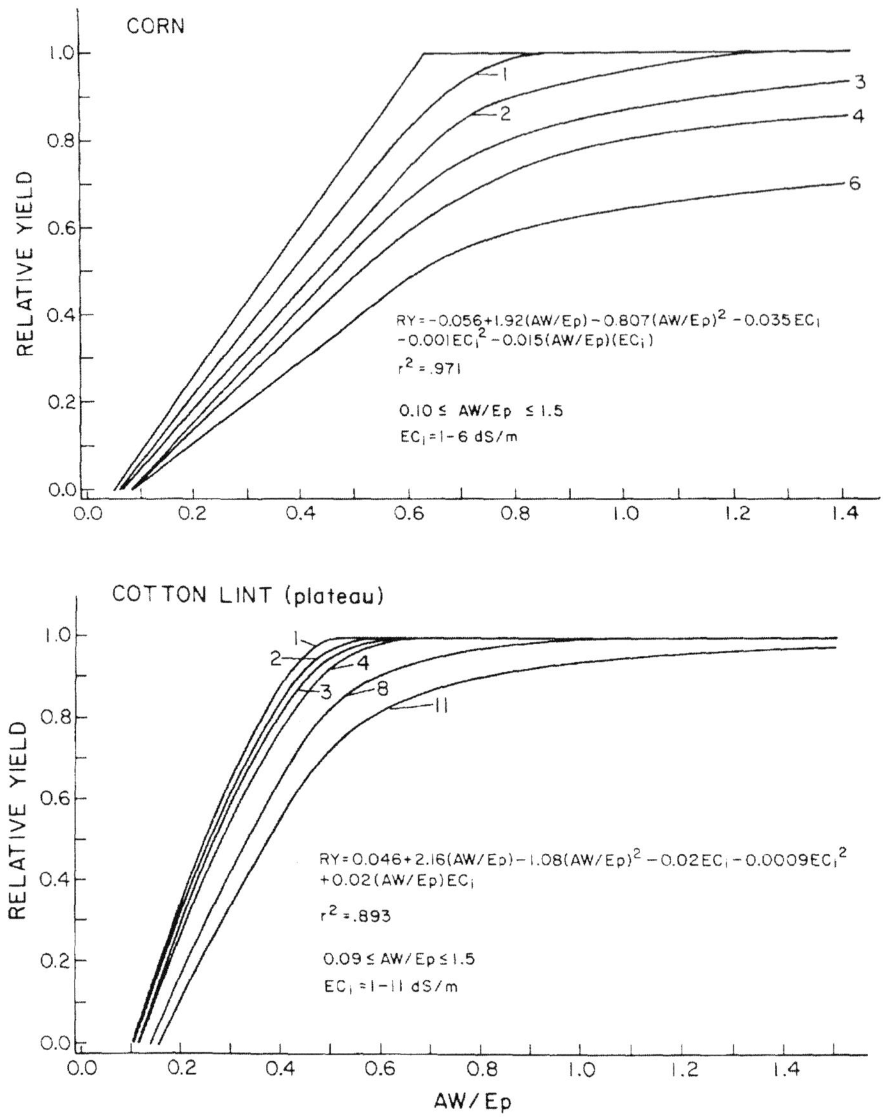

FIGURE 2 Relationships between relative yield of corn and cotton and the amount of infiltrated waters of different salinities. Numbers on curves refer to electrical conductivity (EC) of irrigation water, AW is the amount of infiltrated (applied) water, and E_P represents pan evaporation.

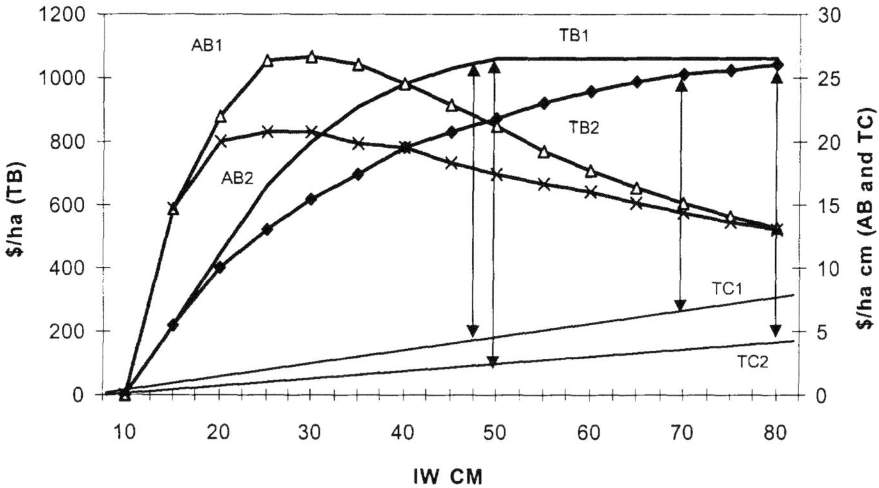

FIGURE 3 The total benefit (TB) and average benefit (AB) as a function of infiltrated water (IW) for two different levels of irrigation uniformity, when irrigation uniformity is more uniform for case 1 than case 2. Also presented is the total cost for water (TC) as a function of IW when the cost for water is higher for case 1 than case 2. The arrows represent the IW that maximizes profits.

The highest profit is achieved where the difference between TB and TC is the greatest. These points are identified by arrows in Figure 3. Some general conclusions can be derived from the information depicted in Figure 3. The economically optimal (profit maximizing) level of IW depends on the shape of the crop-water production function and the price of water. Improving the uniformity of irrigation results in a decrease in the value of IW to achieve economic efficiency. Also, raising the price of water lowers the value of IW to achieve economic efficiency. However, note that raising the price of water has a greater effect on decreasing the economically efficient IW value under the nonuniform irrigation system as compared to the more uniform irrigation system. Indeed, raising the price of water had relatively little effect on changing the economically efficient level of IW for the most uniform system.

A shift in technology or management to achieve more uniform irrigation usually imposes a cost. The increased cost must be offset by the increased benefits associated with improved irrigation uniformity to justify the investment. This factor must be evaluated on a case-by-case basis.

The economically optimal irrigation under nonuniform irrigation results in much more deep percolation than for the more uniform irrigation. Unless the benefits or costs associated with deep percolation are borne by the farmer, they become externalities. With externalities, the economically optimal management from the perspective of the farmer may not be economically optimal from a social perspective. Because externalities can be either positive or negative, and the magnitude varies based on individual situations, it is not possible to make general statements concerning the comparative economic efficiencies of the farmer vs. society in general.

Letey et al. (1990) did an economic analysis of irrigation systems for cotton production in the western San Joaquin Valley of California, where high water tables require the installation of a drainage system. In this case the drainage waters are highly concentrated with salts and selenium, and there is a cost associated with their disposal. If the cost for drainage water disposal was not borne by the farmers, the cheaper nonuniform irrigations were determined to be most profitable for the farmer (Figure 4). However, if there was a constraint on the amount of drainage water generated, or a cost for disposal imposed on the farmer, investment in more expensive uniform irrigation systems was justified.

The average benefit (AB) as a function of IW is also depicted in Figure 3 for the two irrigation uniformities. AB is calculated as TB divided by IW and has the units of \$ (ha cm)$^{-1}$. AB is the average dollar return per hectare-centimeter

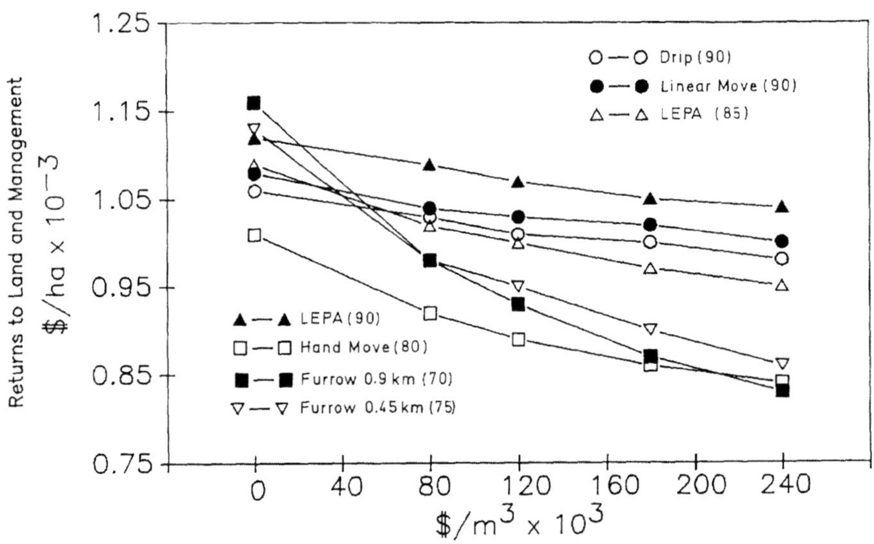

FIGURE 4 Drainage disposal cost effect on returns to land and management for several irrigation systems: hand move sprinkler, linear move sprinkler, low energy precise application (LEPA), furrow, and drip. Numbers in parenthesis for each irrigation system are the assumed Christiansen uniformity coefficient (CUC).

(ha cm) of IW and is not affected by the cost of water. Note that the maximum AB value occurs at a lower IW value than the economically optimal quantity. Even though the average return decreases, the cost for additional irrigation water is exceeded by additional returns associated with an incremental increase in water. The economically optimal input is at the point where the marginal benefit equals the marginal cost. The marginal benefit is defined as the benefit associated with the next incremental increase in input, and likewise the marginal cost is the increase in cost associated with the next incremental input. The marginal benefit is the slope of the TB curve, and the marginal cost is the slope of the TC curve depicted in Figure 3. Note in Figure 3 that the slope of the TB curve equals the slope of the TC curve for the optimal irrigation.

The shape of the AB curves in Figure 3 is identical to the shape of the ratio of yield (Y) to infiltrated water (IW). This ratio (Y/IW) is one definition of water use efficiency. Note that the IW that achieves maximum water use efficiency by that definition is not the IW that is economically efficient. Irrigating to achieve the maximum commonly defined water use results in a significant reduction in yield. Clearly, maximizing water use efficiency by this traditional definition can lead to results that are not the most economically beneficial.

A shift in production function from less uniform to more uniform irrigation does result in a higher water use efficiency for a given value of IW. Therefore, increasing the numerical value of water use efficiency by a change in management that entails a change in production function is positive. However, it is not necessarily economical. It is not obvious that the shift in production function from the less uniform to more uniform irrigation is economically efficient. The main conclusion is that generalizations cannot be made and each situation has to be thoroughly evaluated from production and economic considerations.

IRRIGATION SCHEDULING

Irrigation scheduling refers to the time duration and quantity of an irrigation. Although the crop water production functions depicted in the Figures 1-2 provide the scientific and economic basis for optimizing irrigation, farmers do not necessarily have such complete detailed information available to guide their irrigation management. Nevertheless, the general principles still apply. Since the purpose of irrigation is to replace the water lost from the storage zone between irrigations, knowing the amount of ET that has occurred since the last irrigation is important. Alternatively, the farmer could monitor the soil water content as a function of time to determine when the soil is sufficiently dry to warrant recharge. Therefore, a method of monitoring either ET or the change in soil water content is required for irrigation scheduling.

Climatic conditions drive ET. Therefore, monitoring the potential ET by an evaporation pan, or determining potential ET from other climatological data, is required. In California, several weather stations have been established throughout the state to form a California Irrigation Management Information System (CIMIS). Farmers with computer systems can get daily information from the weather station located nearest to their farms.

The climatological data identify the potential ET that can occur. Crop ET is not always equal to potential ET. For example, during the early part of the season for row crops the plant is small and the crop ET is much less than the potential ET. As the crop grows and its canopy cover increases, the crop ET approaches potential ET. Thus, to estimate the crop ET the potential ET is multiplied by a crop coefficient (K_c), which must be empirically determined for each crop as a function of time. Studies have established the crop coefficients for several crops in California. Results from these studies can be used to guide irrigation management. Nevertheless, the study results are not absolute, and the farmer must use judgment and make observations in the field to assure that his irrigation is appropriate.

Monitoring the soil water content as a function of time requires instrumentation. The neutron probe can be used to measure the water content in a soil profile, but this method is labor intensive. It requires the installation of neutron probe access tubes and then measurements on some predetermined schedule. Other instruments, such as tensiometers, can be installed at various depths and require reading on a timely basis. Some of the instruments have electrical signals that can be connected to a recorder for continuous monitoring with minimal labor input. Soil moisture monitoring requires capital investment and then some level of operational expense.

WATER REUSE

Water reuse implies that the irrigation water had previously been used. It may have been used in an urban setting and discharged as wastewater from a sewage treatment plant, or it may have been previously used for irrigation and discharged as drainage waters. The important factor in determining its suitability for irrigation is the chemical and biological composition of the water and not the fact that it had previously been used. Agricultural drainage waters and sewage waters generally have increased salinity levels that must be considered in irrigation. However, the basic principles presented earlier for using waters of different salinity apply under these conditions.

Sewage waters may have health implications. Regulations on reuse of water for protecting health will be specified by public health agencies and dictate when and how these waters may be used for irrigation. However, they do not affect the basic irrigation principles that have been presented. An irrigator simply needs to consider the chemical composition of the waters, along with the other basic irrigation principles, in guiding his use of reclaimed water for irrigation.

CONCLUSIONS

Many factors contribute to the selection of the irrigation system and management. The scientific and economic principles are well established and can be used to select the optimal set of options. However, the optimal solution is case-specific and cannot be specified in a general manner.

REFERENCES

Letey, J., A. Dinar, and K.C. Knapp. 1985. Crop-water production function model for saline irrigation waters. Soil Sci. Soc. Am. J. 49:1005-1009.

Letey, J., A. Dinar, C. Woodring, and J.D. Oster. 1990. An economic analysis of irrigation systems. Irrig. Sci. 11:37-43.

Status of Agricultural Water Use in Iran

Amin Alizadeh and Abbas Keshavarz

INTRODUCTION

Because water is a critical natural resource, it has always played a vital role in progress and development. Since the first known human empire was established thousands of years ago in the southwestern part of Iran, water has played a key role in any social changes that have taken place in this country. For many people viewing Iran from outside of the country, water scarcity in this country may not appear to be as serious as in other countries of the Middle East. Nevertheless, with a population of more than 65 million people, Iran is actually one of the driest countries of the world. Even if all the renewable water resources could be utilized, excluding the incoming international river, the total amount is not more than 117 billion cubic meters (bcm). Considering that about 88 bcm are currently used each year, the country is left with about 30 bcm of additional water capacity for future use.

Today, the consequence of rapid population increase and the immigration of millions of Afghans is increased pressure for rapid water and land development. In addition, the processes of urbanization and industrialization and the development of irrigated agriculture to support population growth have raised the demand for water, but at the same time have reduced the supply.

Iran is an oil rich country, but water, unlike petroleum, has no substitutes and cannot be purchased in a world market that has many alternative suppliers. If the problem of water scarcity, not only in Iran, but also in other countries of the Middle East, is not solved, its most obvious consequence will be that millions of the people of these countries will seek refuge in other nations. Therefore, water scarcity is not an isolated national problem, but rather is a common problem of all the countries of the Middle East. In order to master the complex problem of water scarcity there will need to be regional cooperation and assistance.

GEOGRAPHICAL LOCATION AND POPULATION

Iran, with an area of 165 million hectares (Mha), is located in a semiarid region of the Middle East. Distribution of precipitation is uneven. The average amount of precipitation over the country is 252 mm/year, which is less than one-third of the world average. While annual precipitation usually exceeds 2,000 mm in some of the northern parts of the country, it may be less than 20 mm in desert areas. Although water surpluses exist in the mountain regions, the areas of high population concentration and high water demand are hundreds of miles away.

Population growth in Iran is high. The highest recorded rate of 3.9 percent occurred in 1986. But a remarkable achievement of Iran in applying family planning programs during the years of 1986-1996 contributed to a lower rate of population growth of 1.45 percent in that decade (Ghazi, 2002). The latest census figures showed the population of Iran to be 60 million in 1996. Today, it is estimated that the population of the country may be more than 65 million. It is also expected that the population may double by 2021 (Plan and Budget Organization, 1999). Rapid population growth in the last two decades has changed the relative composition of the rural and urban populations. While the ratio of rural to urban population was 40/60 before the revolution, it is now reversed. By 2010 some 80 percent of the total population may live in urban areas and especially in big cities like Tehran, Mashhad, and Isfahan. Most of the water resources that sometime ago were used for agriculture are now used to supply drinking water to these cities. Altogether, population growth, urban and industrial growth, and agricultural development in Iran have created a condition of water stress. This situation is beyond a water shortage or crisis and aggregates the serious scientific, technical, ecological, economic, and social issues surrounding water for now and the years to come (Ghazi, 2002).

WATER RESOURCES AND HYDROLOGY

Available data for Iran's freshwater resources are presented in Table 1. As is seen in this table, the average renewable water in Iran is 130 billion cubic meters only. However, since the hydrologists have been involved in measuring rainfall and river flow at catchment scales, these data may not give a complete and accurate image of water availability unless data are also measured at other scales.

The level of water stress depends upon technical scarcity, demographic scarcity, and hydraulic density of population (Falkenmark, 1999). Given the high population increase and recent persistent drought conditions, Iran's average annual supply of renewable freshwater per person fell from 2,254 m^3 in 1988 to 1,950 in 1994, and the estimated figures for the years 2005 and 2020 are 1,750

TABLE 1 Water Availability and Use in Iran

Component	Volume (bcm)	Percent of Total
Precipitation	413	100
Evaporation	283	70
Renewable water	130	30
Surface water	105	
Groundwater	25	
Total water use	87.5	100
Agriculture	82.0	94.25
Domestic	4.7	4.75
Industry (etc.)	0.8	1

and 1,300 m^3, respectively (Ghazi, 2002). Biswas (1998) believes that generally a country will experience periodic water stress when freshwater supplies fall below 1,700 m^3 per person per year. Given this statement, Iran is beginning to encounter water stress. Based on the data in Table 1, almost 70 percent of all annual freshwater resources in Iran are already under use, and the remaining 30 percent may not be technically feasible to use. As far as hydraulic density of population is concerned, spatial distribution of water resources in Iran is uneven. Almost 30 percent of all annual freshwater of Iran is concentrated at the south-western part of the country, where only a very small percent of the population is located. According to these figures, and based on available freshwater resources, the population of Iran has reached its maximum capacity unless sustainable policies are focused on demand management.

IRRIGATED AGRICULTURE

According to the figures in Table 1, more than 94 percent of the total annual water consumption in Iran is used for agriculture, whereas the percentages for domestic and industrial uses are 4.75 and 1 percent, respectively. Even though most of the renewable water is used in agriculture, the productivity of water (ratio of yield per unit of water) is very low. Out of all water used in agriculture, 50 to 60 million tons of food material are produced. Therefore, the economic value per cubic meter is 0.75 kg/m^3. The economic value of agricultural products in Iran (including rain-fed agriculture) is estimated to be U.S. $4.75 billion, which is about 26 percent of the gross domestic product (GDP). Despite the fact that only 10 percent of the total area of the country is arable, the area under irrigation is not more than 30 percent of the total cultivated land. Over the last decades, many attempts have been made to increase the area under irrigation by supplying water through construction of dams, but these attempts were not significantly effective. The amount of water regulated by dams in Iran is estimated to be 25 billion m^3, which is 28 percent of the surface water used in agriculture (Plan and Budget Organization, 1999). As has been reported, no increase in irrigated land has been seen during the last four decades, except some development during the period of 1980-1990 (Ghazi, 2002). Although more groundwater was exploited and seven large dams were constructed during the years 1990 to 2000, the irrigated land area did not increase. While in 1999 Iran was the ninth highest wheat importer and the third highest rice importer in the world, in the year 2001, due to a prolonged drought, Iran rose to the position of fifth-highest wheat importer in the world. It should be remembered that in spite of heavy storage of oil, Iran is still an agricultural country, and the agricultural sector plays a very important role in the national economy. Almost 27 percent of the gross national product (GNP) and 23 percent of labor forces belong to agriculture (Plan and Budget Organization, 1999). It is predicted that a decade from now (around 2012), in order to feed the population of the country more than 100 million tons of food will be needed. By that time, water demand will increase to 126 bcm, and by the year 2021 it will exceed 150 bcm; this is 15 percent in excess of the country's total potential renewable freshwater resources.

WATER USE EFFICIENCY

Considering that 98 percent of all agricultural raw materials in Iran are produced from irrigated lands it must be admitted that theoretically this is not a difficult problem to solve. Agricultural activities account for 94 percent of all water used in Iran. Since water use in the agricultural sector is inefficient, considerable improvement is possible through policy changes, technological solutions, and other alternatives. The situation is further complicated by the fact that agriculture is the major economic sector in Iran, and it will not be an easy task to convince the farmers of the need for water pricing policies to improve agricultural water management. Therefore, water management and development will be an increasingly challenging task in the near future. This is the main reason why more scientific cooperation and exchange of ideas and experiences is needed.

PROBLEMS OF IRRIGATION AND DRAINAGE NETWORKS

One of the problems of water management in Iran is the way that we view our irrigation and drainage projects. This vision does not coincide with the farmers' indigenous knowledge of agriculture. The dry, high desert climate in the plateau of Iran and the scarcity of water resources in the area have forced farmers to develop special methods of using their limited natural resources with maximum care. They have tried to manage water and water resources using an understanding of nature that they have acquired through experience. The harmony they have established through the ages with natural processes has brought them a peaceful life through which they have been able to satisfy their needs despite limited resources, an inhospitable environment, and inadequate exploitation methods. Their struggle against nature has led to the accumulation of an indigenous knowledge base for conducting life and managing the environment. This indigenous knowledge has then served as a launching pad for further experience and acquaintance with easier and more efficient methods. Water exploitation methods are another facet of their life. Over the years these methods have evolved into what are known as current irrigation practices. These practices are the result of customs, practices, beliefs, knowledge, and experience that they have learned and put to use generation after generation. Unfortunately, the unilateral governmental vision to implement and manage irrigation and drainage projects in the past has been solely physical, and the participation of farmers or nongovernmental organizations has been neglected. The results of such a one-dimensional vision have been the disassociation of cultural and social relations between farmers and their system of irrigation.

Even the physical development of irrigation and drainage projects has not been completed. Today, of the 26 bcm of water that are stored in reservoirs, only 17 bcm of water are used for agriculture, while 9 bcm are released just to operate hydropower units. Although the area of agricultural lands below the dams is 1.7 Mha, only 1.2 Mha of project area are equipped with main irrigation canals and 0.5 Mha are equipped with distribution canals.

IRRIGATION TECHNOLOGY IN IRAN

Operationally, the technology that has been adopted for new irrigation projects is not appropriate. It is true that the irrigation industry is not supposed to do original research and produce original technology. Rather, we look to the experience in other parts of the world and attempt to use good judgment and management, guided by the public interest, to transfer and apply the best available and most appropriate technologies.

If technology transfer does not take place with a view to provisions for future developments and to the characteristics of our societies, it is doomed to cause irreparable damage. Such problems are now seen in the modern irrigation projects that have been constructed in Iran during the past 20 to 30 years. Developing countries are suffering from the consequences of improper use of technology more than developed countries despite the fact that developing countries are expected to have learned from the experience in industrial countries. The situation in developing countries stems from the fact that modern technology originated outside their boundaries, and with the import of technology the indigenous social system has also been intruded upon. Every technological import from industrial countries to developing countries has associated with it some unilateral colonialist interests for the exporter, but such technology transfers are more successful if the interests of both parties are secured. What this means is that all aspects must be considered in any case of technology transfer.

Since technology is usually transferred to developing countries from outside, these countries can face more problems in providing the appropriate social changes to accompany technological developments than the industrial countries in which social development takes place simultaneously with technological development. In developing countries this integration takes time and money. It has been argued by many that if technology transfer is uncontrolled and independent of objectives or values, it will affect all aspects of our lives and sometimes make us slaves of technology rather than vice versa.

A careful examination of the relationships between farmers and imported technology in modern irrigation projects of Iran reveals that when farmers have kept a safe distance from technology, they have maintained for themselves the right to choose an appropriate technology that enhances their goals. However, where this distance has disappeared, they have failed to exercise any control over their decisions, leaving them ultimately only the choice to either give up technology altogether or to follow not a progressive, but rather regressive path as dictated by technology. There have been some cases in the past in which an irrigation project has been constructed with a modern system of water distribution. But when the project was given to local farmers, some feature of the project was changed by them, because the new system was not appropriate to them, negating some of the intended benefits of the project. To avoid such catastrophic choices, technology and its relevance to each particular situation must be studied and then an appropriate form of technology must be introduced.

APPROPRIATE TECHNOLOGY

Over the past decade, technology has been increasingly in demand as a means of development. Particular technologies may be obtained in one of the following three ways:

1. as finished technology from other countries,
2. as imported technology but adapted to domestic needs, and
3. as indigenous technology.

In most cases a technology which requires a huge capital investment is cheaper to import than to develop, but the imported technology may not be fully applicable without some modification. Therefore, there is usually a mediating investment involved to balance or adapt the imported technology. This will increase its integration and applicability with the domestic situation. In cases where importing such technologies is not possible, it must be developed within the country based on indigenous knowledge. Indigenous knowledge is the integration of all accumulated personal, social, and historical experience. Indigenous knowledge concerning irrigation practices has accumulated and has been applied over thousands of years of experience in Iran. The study of indigenous knowledge and understanding of both positive and negative aspects of the relationship between humankind and nature can help us in the design and implementation of developing new projects.

STATUS OF GROUNDWATER RESOURCES

Groundwater is one of the most important water resources of Iran. One of the best methods of supplying water is digging qanats, a practice with a long tradition in Iran. Researchers have considered qanats to be an innovation developed by Iranians about three thousand years ago. A qanat initially consists of a well dug in a mountainside to reach the groundwater stored there. An underground tunnel is then dug from this point, directing the water to the village. Along the way to the village, some access wells are also dug at certain intervals, to provide access for later repairs and cleaning of the tunnel. Some of the main wells of the qanats systems in eastern Iran are more than 400 meters in depth, deep enough to hide the Eiffel tower. Their tunnels are longer than the equator. A great amount of water in Iran is supplied by qanats whose total length is estimated to be 160,000 kilometers. Unfortunately, most of the qanats have become dry due to exploitation of groundwater by pumps and wells.

Groundwater balance shows that there is a difference of 4.8 bcm between recharge of groundwater resources (56.5 bcm) and discharges from them (61.3 bcm). The effect of this unbalance is evident in most of the valleys. Land subsidence, salt intrusion, and lowering of the water table are among the most prominent effects. The average drawdown of the water table in 168 valleys of the country, from which 73 percent of all withdrawals occur, is more than 1 meter per year. In some of the eastern provinces, more than half of the groundwater storage has been depleted.

IRRIGATION EFFICIENCY

A common perception is that irrigation wastes enormous amounts of water in Iran. It is commonly said that if irrigation could just be more efficient, water would be made available for more agriculture and other water uses, and there would not be as great a need to develop more water infrastructure. Unfortunately, this perception is in many cases not true, and the opportunity for real water saving through increased irrigation efficiency is much less than perceived.

It is claimed that the overall irrigation efficiency in Iran is 30 to 35 percent. Briefly, the term "irrigation efficiency" (Ei) is used to define the amount of water that needs to be supplied from a source and delivered to the field (D) to meet water needs of crops (ET). Therefore, irrigation efficiency is calculated as $Ei = ET/D$. However, the data show that this may not be an accurate estimate. According to published agricultural statistics, the area under irrigation in Iran is 7.80 Mha. Of this total irrigated land area, 3.0 Mha are devoted to cereal, 2.0 Mha to orchards, and 2.8 Mha to different field crops. For irrigating these areas, 83 bcm of water is used. In comparison, it is reported that the average net irrigation requirements in Iran for cereal and field crops are 5,100 and 8,100 cubic meters per hectare (m^3/ha), respectively. The Ministry of Energy, which is responsible for water allocation in Iran, has estimated the average amount of irrigation required to be 5,200 cubic meters per ha. The average of figures published by different consulting engineers is 5,900 m^3/ha. Considering these figures, the overall irrigation efficiency in Iran will be somewhere between 48 and 55 percent, which is quite different from the figures that are presented officially or unofficially by various sources. Since the above figures are taken from various sources, we have to change our vision toward irrigation efficiency. If data for the area of irrigated lands are correct, then with an irrigation efficiency of 35 percent at least 120 bcm of water would be used, rather than the reported 83 bcm. If the reported volume of 83 bcm currently used for agriculture is right, then with an irrigation efficiency of 35 percent, the area under cultivation would be about 5.2 Mha rather than the reported 7.8 Mha.

Based on the available data, the area that receives full irrigation in Iran is 2.5 Mha. Half of this area is equipped with modern systems of irrigation and is operated by government organizations. Irrigation efficiency in such systems is very low and is measured to be 20-30 percent. The reason for such low efficiency may be due to the free availability of water released from dams, giving no incentive to farmers to save water. The other half is operated by the private sector, and the water is supplied from groundwater resources. Here, also, the irrigation efficiency is rather low and has been measured to be about 35 percent. The rest of the irrigated farms in Iran belong to small farm holders who do not intentionally save water, but their irrigation efficiency is quite high. Irrigation efficiency in these farms is estimated to be 55-65 percent. The reason may be due to deficit irrigation, which they usually practice. These farmers have enough land, and the area under their cultivation is much greater than the water available to them. They usually get more benefit from extensive farming with deficit irrigation compared to intensive farming and full irrigation. Considering the above situation, the area under cultivation is balanced with the amount of water supply and demand.

MAJOR PATHS TO SAVE WATER

While the real irrigation efficiency in Iran is not low, many experts still believe enormous amounts of water are wastefully used in agriculture. Perhaps the major reason for the discrepancy is the way that the word "efficiency" is understood in irrigation. This efficiency may vary between 90 percent in the case of drip irrigation and 20 percent in traditional paddy irrigation systems. Thus, it is reasonable to think that increasing irrigation efficiency can save a large amount of water. In many cases it is possible but in other cases it is not. This possibility to increase irrigation efficiency depends on what happens to drainage water that is delivered to the field but not used by the crop for evapotranspiration. Drainage water may flow to the sea or desert, in which case it is effectively lost to further use. In this case, increasing irrigation efficiency will result in real water saving. On the other hand, drainage water may flow to other surface or subsurface areas where it can be captured and beneficially reused. This is what is called the return flow of water. Return flow may be a major source of recharge for aquifers in irrigated areas. One person's drainage may be another person's water supply. Thus, countrywide it is advisable to think in terms of basin efficiency rather than field efficiency, as the basin efficiency number is probably a better indicatior in terms of real water savings that can be made.

By introducing the concept of basin efficiency, the effect of return flows is taken into account. In Iran, although the typical efficiency in full irrigation farms is 35 percent and in deficit irrigation farms is 55 percent, for the Iranian irrigation sector as a whole, at basin level it is high due to recycling. All things considered, there is not much real water saving to be made through improved irrigation efficiency in Iran, even though the system appears to be inefficient at first glance. The problem with the concept of increasing efficiency, even considering basin efficiency, is that the value of water is considered only in terms of physical quantities. Irrigation efficiency does not fully capture the value of water. This is the subject of the more general concept of increasing water productivity.

AGRICULTURAL WATER PRODUCTIVITY

Water productivity is simply defined as the amount of production per unit of water applied in the field. Water productivity can be increased by obtaining greater production with the same amount of water or by reallocating water from lower to higher value crops or from agriculture to other sectors where the marginal value of water is higher. Indeed, the greatest increases in the productivity of water in irrigation have not been from better irrigation systems but rather from increased crop yields due to better management. For this reason, it is generally best to use the term water productivity, rather than efficiency.

A key to mitigating the problem of water scarcity in Iran is increasing the productivity of water. Let us consider agriculture, as there is a tremendous need to produce food, mainly grains, from water resources. There are generally four paths of generating more agricultural output from national-level utilizable water resources as follows:

1. increase utilizable water,
2. develop more primary water (increase in development of facilities),
3. consume more of the developed water beneficially (increase in basin efficiency), and
4. produce more output per unit of water consumed (increase in water productivity).

Opportunities to increase agricultural water productivity via the first two approaches are very limited in Iran. If we do not consider the incoming international water, the total amount of renewable water will not exceed 117 bcm, from which only 105 bcm can be controlled as utilizable water. For the third approach, it may be feasible to improve water conservation in those areas that receive full irrigation. By proper water management and improvements in water distribution networks, irrigation efficiency may increase up to 50 or even 60 percent. As a result, the area of these fully irrigated lands may increase from 2.5 Mha to 3.5 or 4.3 Mha, respectively. Although it is also possible to increase irrigation effectiveness in those areas that receive deficit irrigation, no increase in cultivated area could be expected. Thus, the major solution will be the fourth approach, increasing water productivity.

In agriculture, water is consumed as evapotranspiration, which is essential for crop production. As Smith (1999) stated, the process of converting water into yield is referred to as the eco-physiological water use efficiency or crop water productivity (CWP). Crop water productivity can be defined as the kilograms (kg) of yield per unit of consumed water (ET), or CWP = kg / ET. An extensive data set of studies carried out worldwide has shown that global values for CWP are very low for some crops (Doorenbos and Kassam, 1979) and generally high for some other crops like tomatoes. The genetic characteristics of the crop are the primary factors determining CWP. A secondary factor that in various ways affects CWP is the reaction of the crop to water stress.

For cereal grains, water productivity ranges between 0.2 and 1.5 kg (grain)/ m^3(water) (Molden et al., 1999). Based on area under cultivation and the amount of grain yield in the Khorasan Province of Iran, water productivity for this crop is about 0.7 kg/m^3, which is quite low compared to a similar environment like the Imperial Valley of California. The Imperial Valley is situated in a desert environment like Khorasan Province. The range in wheat yield in these areas is from 2 to 6 tons per hectare. Within the Khorasan Province, there is a great variability in yield, with some farmers achieving productivity levels as high as those in California and some farmers well below the average. Of course, production levels also depend on environment, market, soil, and other conditions that are not identical in any two sites. In spite of these differences, there appear to be opportunities to manage resources to achieve greater productivity. As a rule of thumb, a reasonable level of water productivity for wheat is about 1.0 kg/ m^3. Therefore, if the demand for grain grows by 50 percent in the country by 2020, one way to match this increase is to increase water productivity by 50 percent.

Water use efficiency is a combination of water productivity, which is a biological factor, and irrigation efficiency, which is a physical factor. Therefore, the menu of options for improving water use efficiency includes a combination of measures in four different areas as follows:

1. Technical practices, such as land leveling for uniform distribution of water on land.
2. Institutional practices, such as establishing a water user organization, reducing irrigation subsidies, and fostering rural infrastructure for private sector dissemination of appropriate, efficient technology.
3. Agronomic practices, such as developing varieties of crops that yield more mass per unit transpiration, substituting crops that consume less water for crops that consume more water, intercropping to maximize use of soil moisture, and selecting drought tolerant crops where water is scarce.
4. Managerial practices, such as better irrigation scheduling, recycling of drainage and tail water, and precision irrigation.

Precision irrigation refers to a combination of technologies and associated management practices that can help overcome the problems of quantity and timing of irrigation supply. For example, drip, sprinkler, and level basin techniques enable control of water applications and can contribute to achieving higher crop yield. The results of experiments carried out at Khorasan Agricultural Experiment Station on melon, tomato, sugar beet and watermelon showed that precise drip irrigation can increase water use efficiency up to three times.

It should be remembered that precision irrigation does not imply expensive high technology irrigation. Instead, it refers to a broad range of technologies and water management practices that enable farmers with limited access to water to apply water to their crops in the time and quantity that increase the productivity of water. Precision irrigation can even be practiced with existing conventional technologies. Recent development of a mechanized pot system of irrigation in Iran has shown that this system tremendously reduces irrigation requirements, and it successfully has been used for irrigation of orchards (such as pistachio), vegetables, and green houseplants.

SELECTED BIBLIOGRAPHY

Alizadeh, A., G.H. Kamali, and I. Malek-Mohammadi. 2001. Sustainable utilization of saline water in agriculture based on indigenous knowledge. International symposium on prospects of saline agriculture in the GCC countries, Dubai, United Arab Emirates.

Biswas, A.K. 1998. Deafness to global water crises. Causes and risks. Ambio 27(6):93.

Doorenbos, J. and A. Kassam. 1979. Yield response to water. Rome: FAO Irrigation and Drainage Paper 33, p. 193.

Falkenmark, M. 1999. Forward to the future. A conceptual framework for water dependence. Ambio 25(3):211-212.

Ghazi, I. 2002. Water resources management and planning in Iran. Report to the University of Isfahan.

Molden, D. and C. de Fraiture. 1998. Major path to increasing the productivity of irrigation water. Chapter 4, Draft 14-5-98. International Water Management Institute (IWMI), Sri Lanka.

Plan and Budget Organization, 1999. Document of the Third Plan, Volume 1. Iran Ministry of Water and Power, Tehran.

Reza, A., Kooros, Ch., Emam-Shooshtari, M., and A. Entezami, 1973. Water and irrigation technique in ancient Iran. Ministry of Water and Power, Tehran.

Sally, H., Sakthivadivel, R. and D. Molden, 1999. More crop per drop. 6th International Micro-irrigation Congress, Cape Town.

Seckler, D., A. Marasinghe, U. Molden, D. de Silva, R. and R. Barker, 1998. World water demand and supply, 1990 to 2025: Scenarios and issues. IWMI, Sri Lanka.

Smith, M. 1999. Optimizing crop production and crop water management under reduced water supply. 6th International Micro-irrigation Congress, Cape Town.

Agricultural Drought Management in Iran

Sayed-Farhad Mousavi

INTRODUCTION

Deserts and arid lands have always presented a harsh environment in which to live, and their sunburned people make a great effort to survive. They have done their best to find ways of better living. Water is a vital commodity in these areas, and the art and technology of saving water for consumption is critical to these people.

Drought is a natural phenomenon that can occur in any region and cause economic, social, and environmental losses. It has greater effects in dry and water-deficient regions and is a global threat, but it has not been studied in detail in many countries. During recent decades, drought has exceeded any other natural disaster in number and frequency.

Drought is the result of an abnormally dry period that lasts long enough to cause an imbalance in hydrologic processes (storage and consumption). Reduced precipitation or increased temperature, either alone or together, can cause drought. At the present time, 40 percent of the population of the world is confronted with periodic droughts, affecting some arid and semiarid countries in the north of Africa, parts of India, the north of China, the Middle East, Mexico, Middle Asia, Australia, and the western United States. Severe climatic changes, increase of greenhouse gases, and El Niño/Southern Oscillation (ENSO) phenomena are responsible for these droughts. When vegetation cover is destroyed, droughts are intensified. Climatic, hydrologic, and agricultural droughts are the most important categories.

WATER RESOURCES OF IRAN

The state of water resources in Iran is summarized as follows. The main source of water is precipitation, which normally amounts to 252 mm or 413 billion cubic meters (bcm) annually. This precipitation depth is less than one-third of worldwide average precipitation (831mm) and about one-third of the average precipitation in Asia (732mm). About 30 percent of the precipitation is in the form of snow, and the rest is rain and other forms of precipitation. While 1 percent of the world population lives in Iran, our share of renewable freshwater is only 0.36 percent. Of the 413 bcm of annual precipitation, 296 bcm are lost as evapotranspiration, 92 bcm runs as surface flows, and 25 bcm infiltrates into groundwater resources. Annually, about 13 bcm of water flows into Iran from neighboring countries. So, total renewable water resources are 130 bcm annually. From these sources, about 88.5 bcm is withdrawn, of which 82.5 bcm (93.2 percent) goes to agriculture, 4.5 bcm (5.1 percent) is for drinking, and 1.5 bcm (1.7 percent) is allocated for industry, mines, and miscellaneous uses. While the world uses 45 percent of its freshwater resources, Iran uses about 66 percent.

Precipitation in Iran does not have spatial and temporal uniformity. Part of the country receives less than 50 mm, while the northern part receives more than 850 mm of rain annually (Figure 1). More than 50 percent of the rain falls in winter, and less than 18 percent falls in summer. From the middle of the spring, river and stream discharges start to decrease, and groundwater is the only water source for summer and fall seasons. Statistics show that in 1996 and 2000 about 59.41 and 61.2 bcm, respectively, were withdrawn from the aquifers. Nonuniform temporal distribution of precipitation causes droughts in the years when most annual rainfall occurs in a short time and runs off quickly.

On the basis of studies performed by United Nations (UN) experts, the per capita water resources of Iran are projected to be about 726-860 m^3 in 2025, compared with 2,200 m^3 in 1990. Overpopulation in an arid and semiarid country causes diverse problems, including increased demand for scarce water and intensified competition between different sectors (agriculture, human consumption, and industry). Overpopulation in Iran will contribute to the country reaching a state of water crisis before the year 2025. Unplanned and irregular expansion of the main and satellite cities in the past 100 years has increased the population six-fold and contributed to water shortage problems. In the last 40 years, the population of Iran has increased by 45 million people, 30 million of whom have been added in the last 20 years. The water crisis and water scarcity will intensify in the future.

Water balance of many countries is in desperate straits, since aquifers are exploited severely, water is diverted from the agricultural sector to drinking and industrial supplies, and demand for more food and better diets is increasing. So, water is scarce, and as the studies of the International Water Management Institute (IWMI) show, it will get scarcer (Figure 2). The countries in dark grey in

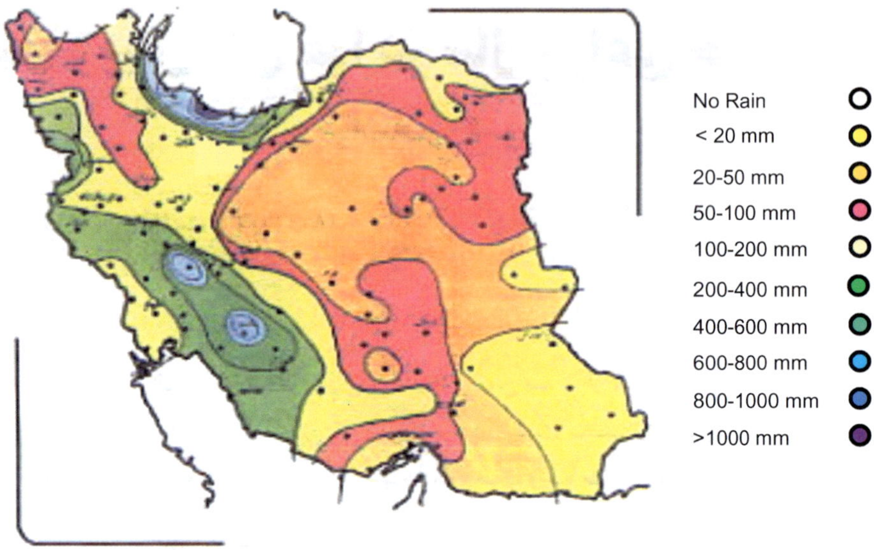

FIGURE 1 Distribution of total annual rainfall (1999-2001).

Figure 2 do not have enough water resources to meet their future demands, despite investment in this section. The countries in medium grey may meet their projected demands, provided new water resources (including large dams) are developed.

Iran is located in an arid region of the earth (25 to 40 degrees north). With more than 90 percent of the country's area in arid and dry regions, drought periods and their effects, which are more frequent than floods, are of great concern to politicians and planners. Drought is a natural and climatologic fact for Iran and should not be considered a surprise. In the last 23 years, there were 13 years of dry periods, and the extent of the social, economic, and human damage was more severe than flood damage. Sensitivity of the southern parts of Iran to severe drought is higher than other parts of the country.

EFFECTS OF DROUGHT

Drought has different direct and indirect effects. The direct effects include decreasing production in agriculture and rangelands, groundwater depletion, low flows in the rivers and streams, exposing of all natural or human ecosystems to destruction and contamination, soil erosion, and mortality of livestock and wild life. Indirect effects include lowering farmers' income, decreasing government's

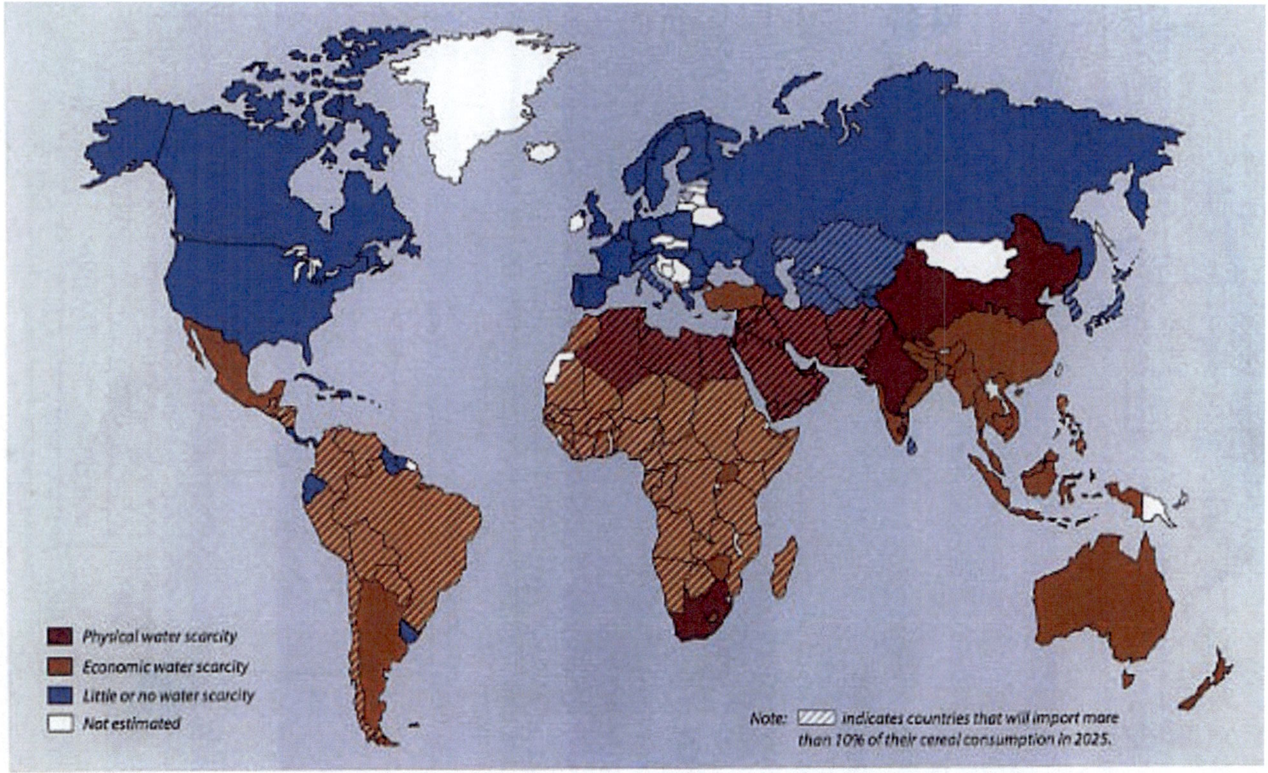

FIGURE 2 Water scarcity projections (Seckler, et al., 1999).

tax income, increasing the cost of water and forage transport, and increasing migration of farmers to small and big cities. Reduced yields and quality of grain, and increased incidence of pests and diseases are the most detrimental effects of drought in the agricultural sector.

DROUGHT DAMAGE

Iran has experienced a severe drought during the last three years (1999-2001), the worst drought period in the last 40 years. On the basis of reports by Iran Meteorological Organization, precipitation during 1999, 2000, and 2001 was 24.2, 40, and 28.5 percent less than normal, respectively. Twenty provinces of the country were faced with the crisis, and 62,500 billion Rials (BR) (U.S. $1 = 8,000 Rials) worth of damage occurred. This damage occurred mainly in the agriculture sector and mainly affected poor people. Drought was more severe in the south, center, southeast, and east. The Khorasan, Sistan and Baluchestan, Fars, Isfahan, Hormozgan, Kerman, Boushehr, Yazd, Ilam, Kermanshah, Lorestan, Markazi, Qom, Hamadan, Semnan, and Tehran provinces were affected the most.

Although parts of the country are confronted with drought in one way or another each year, national figures for damage will not be announced unless it is extensive. A summary of damages that occurred to agriculture, forests, rangelands, and water resources follows.

- The Ministry of Agriculture has announced the damages to farms, orchards, and agricultural structures in 1999 as 10,000 BR. The amount of investment to compensate and mitigate drought effects temporarily was about 27,000 BR (2.5 times 1998 expenditures).
- One hundred cities out of the 700 cities and towns in the country with a population of more than 20 million are confronted with water shortage; more than 20 million head of livestock have been slaughtered or lost; 3,731 springs, 2,500 qanats, and thousands of water wells have dried up. About 200,000 tribal people have lost their only revenue.
- The agricultural sector, which makes up 27 percent of gross national product (GNP), has suffered damage causing GNP to decrease by about 6 percent. The value of 6 million tons of agricultural products is about 8,000 BR.
- Other than vast mining of groundwater resources, with overdraft of 6 bcm per year, 40 percent of the forest land has been converted to other uses in the past 40 years, and forest area has decreased from 21.5 million ha to 12.5 million ha.
- Soil erosion has increased from 10 tons/ha/year in the last decade to more than 30 tons/ha/year, and the total soil loss is about 4 billion tons/year.
- Total surface flows and groundwater recharge of Khorasan Province was 7.5 bcm in 2001, which is 43 percent less than a normal year. Precipitation and river discharge have declined about 40 percent and 90 percent, respectively.

- In Mashhad plain, average well discharge was about 33 liters/second in 1970, but has decreased to 19.5 liters/second in 2001. Electrical conductivity of water has increased 0.035 deciSiemens/meter (dS/m) annually, providing evidence of the increased salinity of groundwater.
- Sistan and Baluchestan, Hormozgan, and Kerman Provinces had 81 percent, 50 percent, and 48 percent rainfall reduction. About 200,000 ha of dryland orchards, more than 10,000 ha of tea farms, and 2,000 ha of banana farms in Sistan and Baluchestan and Jiroft are damaged.
- Overdraft of groundwater resources has caused negative balance of groundwater, and out of 612 plains of the country, 150 plains are restricted.
- More than 70 percent of Fars Province has experienced severe to very severe drought. As a result, 37 percent of the rangelands in this province are damaged 70-100 percent. In 2000, hay production was reduced by 52 percent.
- In Isfahan Province, the extent of damage to 80,000 ha of agricultural land and 79,000 ha of orchards was 2,840 BR. The damage to wells and qanats was 380 BR. The drought of 1999 affecting 6.6 million ha reduced hay and grass production by 20 percent, valued at 24 BR. Drawdown of the water table in wells was 1.5-8.5 m for 1999-2001, although many wells were dry and abandoned completely. In 2000, 762,000 ha of land was not cultivated or had an unsatisfactory yield. Average reservoir storage capacity of Zayandehrud dam decreased by 58 percent. Modern irrigation networks did not receive water or were operating with less than normal capacities.
- Severe droughts caused a reduction of water storage in Sepidrud dam, located in Gilan Province in the north of Iran where droughts are much less common. The farmers were forced to dig wells, and since they did not have much experience with this condition, they were affected badly and sustained a lot of damage.
- In Kohgiloyeh-Boyerahmad Province, grass and hay production from 17,000 km^2 rangelands has been reduced by one-half.
- In Khuzestan province, discharge of the Karkheh, Karun, and Marun Rivers has decreased by 49, 37, and 40 percent, respectively, in 2000-2001 as compared with the average of the past 32 years.
- Displacement costs of wells have not yet been reported but are expected to be high.
- Migration of rural and tribal people and farmers to small towns and big cities has caused numerous problems.
- 6,000 BR of credit was allocated to mitigate drought effects in 2001. Of this amount, 4,000 BR was given to banks as financial aid, 500 BR as financial assistance to farmers, and 1,000 BR to provinces. The share of Khorasan, Fars, Kerman and Sistan, and Baluchestan Provinces was 51.7, 50.8, 45, and 44.3 BR, respectively, of the 1,000 BR assistance to provinces.

MEASURES TO MITIGATE DROUGHT EFFECTS

Water resources experts, agronomists, politicians, and citizens have proposed different measures and approaches to mitigate the adverse effects of agricultural drought. A summary of these proposals is as follows, and the possibility of applying such measures needs further discussion and investigation.

- Use appropriate irrigation methods and technologies (surge, low energy sprinkler, trickle).
- Achieve better allocation of available water. Irrigation systems and farmers should be flexible enough to adapt to changing conditions. Different degrees of scarcity demand different allocation rules.
- Substitute the maxim of "more crop per ha" with "more crop per drop." At the present time, only 0.7 kg is produced from each cubic meter of water. The irrigated land is 8 million ha, with an average yield for wheat, 2,844 kg/ha, cotton, 2,237 kg/ha, sugar beet, 32,224 kg/ha, and citrus, 11,357 kg/ha.
- Practice the "reduce, reuse, and recycle" water-usage maxim in each sector.
- Breed crop varieties to tolerate water deficiency and poor quality irrigation water.
- Substitute "strategic agriculture" with "trading agriculture."
- Prevent surface and groundwater pollution.
- Substitute "crisis management" with "risk management." At the present time, crisis management is applied for curing drought effects and damage.
- Establish a National Drought Commission.
- Establish citizen participation and local management systems.
- Improve watershed and aquifer management. These measures have different benefits such as control and storage of flood flows, groundwater recharge, hay production, increasing vegetation cover, return of farmers to their villages, control of desertification, and employment possibilities. About 41 million ha of alluvium is suitable to receive flood flows, and 50 bcm of floods could be distributed over 15 million ha of alluvial plains.
- Improve collaboration and communication between the research sector and the water custodians.
- Develop local water-harvesting methods and soil moisture storage.
- Conserve forests and rangelands.
- Prevent desertification.
- Educate managers, planners, and producers.
- Encourage foreign investment.
- Conserve the environment.
- Manage water consumption patterns. The percent of people living in urban and rural areas is 63 and 37 percent, respectively. About 32 percent of the water consumed comes from surface and 68 percent from groundwater sources.

- Control population growth.
- Prevent decline of rural and tribal economies.

Water scarcity in the future will mainly be due to unplanned usage and not water shortage. It is not too late to consider the actual facts and figures of Iran's water resources carefully and restructure agricultural drought management. Drought can serve as a catalyst for positive change—a move toward new and more sustainable approaches to managing water. Above-average rainfalls in one or two years should not deceive us into forgetting drought and water scarcity problems. We should think of the future generations.

SELECTED BIBLIOGRAPHY

Bird, J. and P. Wallace. 2001. Dams and development-An insight to the report of the world commission on dams. ICID, J. Irrigation and Drainage 50(1): 53-64.

Dealing with drought. 2002. available at www.cgiar.org/iwmi/home/drought.htm.

Perry, C. 2001. World commission on dams: Implications for food and irrigation. ICID, J. Irrigation and Drainage 50(2): 101-107.

Postel, S. 1999. Pillar of Sands. London: Norton and Company.

Proceedings of First National Conference on Drought Mitigation and Water Shortage, Feb. 27-28, 2001, Jahad Daneshgahi, Shahid Bahonar University, Kerman, Iran.

Proceedings of First National Conference on Water Crisis Mitigation Strategies, March 9-10, 2002, University of Zabul, Zabul, Iran.

Seckler, D., U. Amarasinghe, D. Molden, R. de Silva, and R. Barker. 1999. World water demand and supply, 1990 to 2025: Scenarios and Issues. International Water Management Institute (IWMI).

Water scarcity variations within a country. 2000. A case study of Sri Lanka, available at www.isnar.org.

Wilson, E. 1999. One third of world's population 2.7 billion people will experience severe water scarcity by 2025, says new study. www.futureharvest.org/news.

Coping with Drought

John Letey

GENERAL CONCEPTS

Semiarid regions of the world typically have wide variations in annual precipitation amounts. The maximums and minimums can differ greatly from the long-term average. The term drought here will be used to refer to those times when the precipitation is very low compared to the long-term average. It is not unusual for these very dry periods to span several years.

The severity of the drought impact depends on the hydrologic setting of the locale and previous plans made to accommodate a drought. Surface water supplies are very sensitive to temporal variations in precipitation. Groundwater supplies are moderated by the soil-water flow regime. The magnitude of the temporal oscillation of surface water supply is dampened as the surface water migrates to groundwater. Thus, the impact of a drought is largely dependent on the ratio of available surface water to groundwater supplies for a given region.

Storage enables water saved during wet years to be made available during dry years. Thus, a large storage capacity can moderate the impact of drought. Reservoirs are one commonly recognized form of storage. Groundwater can also be considered to be stored water. In some cases, the full storage capacity for groundwater is not fully exhausted by the natural hydrologic process. Engineering the system to purposely move surface water into subsurface aquifers is a means of increasing storage.

When large amounts of groundwater and other forms of stored water are available, the impacts of drought can be reduced. Nevertheless, under an extended very dry period the normal demand for water cannot be fully accommodated by withdrawing stored water supplies. Therefore, some decrease in demand is required.

The users of water can be broadly categorized into three groups: urban, agricultural, and fish and wildlife. Very efficient water resource management, which is particularly critical during a drought period, depends upon willingness and opportunity for these three user groups to cooperate. Cooperation could entail, among other things, transferring water supplies among user groups. For fish and wildlife, it also might involve having relaxed environmental regulations during dry periods.

DECREASING AGRICULTURAL WATER DEMANDS

Although there are opportunities to decrease demands in all sectors of society, this discussion will only address agriculture. Various options are available for a farmer whose water supplies have been reduced. The options vary depending on whether the farm grows perennial crops, annual crops, or a combination of the two. Annual crops allow the greatest flexibility because they can be selected based on their seasonal water demand. Crops have different seasonal water demands, and those with the least demand can be selected for planting. Another choice is to not plant a crop but leave the land fallow. Depending upon the available water supply, it may be economically beneficial to restrict the acreage planted and irrigate this acreage to achieve high yields rather than to spread the water on the entire farm with yield reductions in all fields.

Perennial crops have a more critical water need than do annual crops. For perennial crops, the highest priority is to provide sufficient water to keep the crop alive until water supplies increase in future years. Because perennial crops have the highest priority and have limited flexibility, the opportunity to transfer water from annual to perennial crops is desirable. Farms that grow mostly annual crops have the opportunity to greatly reduce their water demands and make water available for other users. However, a mechanism for transferring water is required to make this arrangement feasible.

CASE STUDIES

Some case studies from California on coping with drought will be presented. Each of the case studies involves not only agriculture, but also cooperative arrangements between the agricultural and urban sectors. The vast network of canals in California facilitates the transfer of water between various water users from different locations in the state. Some features of these case studies may not be feasible in other geographic locations that do not have water conveyance systems. In addition, the abundance of both surface and groundwater resources that allow conjunctive use programs may not be present in other places.

California State Drought Water Bank

California experienced a drought between 1987 and 1992. An emergency state drought water bank was established in 1991 under the direction of the California Department of Water Resources (DWR). DWR offered to purchase water at a price of $125 per acre-foot from any willing seller and then sell water at the price of $175 per acre-foot plus delivery costs. Under this pricing arrangement, 820,000 acre-feet of water were purchased but only 555,000 acre-feet were sold. The excess 265,000 acre-feet were placed in storage for future years. This storage bank was a regulated water bank in which prices would be established by the market. In 1992 DWR operated another water bank under different conditions. To bring buyers and sellers more into balance, DWR purchased water for $50 per acre-foot and sold it for $72 per acre-foot. In addition, DWR only bought water after a purchaser was identified. However, increased precipitation during that year decreased the severity of the problem and largely negated the need for the bank.

Some general results of the bank experience were that about one half of the water that was purchased came from farmers who fallowed their land. Some water came from increasing groundwater use and decreasing surface water use by farmers. In this case, consideration had to be given to the interconnection between surface and groundwater supplies. In some areas increased groundwater pumping lowered the water table creating a hydraulic gradient that drew more water from rivers. In other words, some groundwater use came at the expense of surface water. Waters were transferred both to the urban and agriculture sectors. It was estimated that the urban areas supplied with water had a $91 million benefit. The agricultural areas to which water was transferred received benefits that exceeded losses in the areas that fallowed their land. Crop sales dropped 20 percent in areas that sold water, but it was estimated that the overall third party impacts were not great. However, the third party impacts appeared to be excessively concentrated in certain areas. For example, Yolo County estimated increased unemployment social service costs of $130,000 that they attributed to farm workers who were unemployed because of the fallowed land. However, there was not sufficient evidence to directly link unemployment to fallowed land.

Arvin-Edison Water Storage District

The Arvin-Edison Water Storage District is composed of approximately 132,000 acres of prime agricultural land located southeast of Bakersfield, California. Approximately 100,000 acres are developed as irrigated vineyards, truck crops, potatoes, cotton, citrus, and orchards. The long-term average rainfall in the district is about 8.2 inches per year and occurs largely during winter and spring months. Therefore, agriculture is almost entirely dependent on irrigation. Prior to the transfer of irrigation water to the district from other areas, the absence of perennial surface streams in the district required that all irrigation water be obtained from groundwater reserves. As a result, considerable groundwater overdraft occurred. The continued lowering of groundwater levels required pumping lifts to exceed 600 feet in many areas of the district. Further, the receding water table, in certain areas, had induced subsurface movement of water with high boron concentrations from the bedrock complex bordering the district to the east into the pumped aquifers underlying the area.

The district's water service contract with the United States Bureau of Reclamation (USBR) provided for delivery of up to 40,000 acre-feet per year of Class 1, or firm water, and up to 311,675 acre-feet per year of Class 2, or non-firm water. During some dry years, less than 40,000 acre-feet of firm water were delivered. Over the long term, the district's annual water entitlement has ranged from a minimum of about 10,000 acre-feet in a very dry year such as 1977 to a maximum of 351,675 acre-feet in very wet years such as 1978 and 1995.

A plan was developed to regulate the erratic water supply to meet an annual irrigation demand schedule. The conceptual plan included a combination of using surface water and groundwater storage. The district has more than one million acre-feet of underground storage capacity to be utilized to regulate the district's water supply.

A key feature of utilizing the underground storage capacity is the district-operated spreading works. Sycamore Spreading Works comprises a total area of 569 acres and is located on the Sycamore Creek alluvial fan near the middle of the district. The Tejon Spreading Works is located on the Tejon Creek alluvial fan and covers an area of 516 acres. A geological fault line acts as a wall that limits the lateral flow of the aquifer and, therefore, retains the recharged waters within the geographic confines of the district.

One operational difficulty of the Arvin-Edison Water Storage District has been to maintain the spreading basin infiltration rates over prolonged periods of spreading. Methods employed to maintain or restore infiltration rates include (1) periodic drying of the surface, (2) promotion of grass growth on pond surfaces, (3) termination of spreading operations when total suspended solids in the imported water exceeds 25 parts per million (ppm), (4) restriction of vehicular travel within the basins, (5) scarification of surface soils by chiseling or disking, or (6) removal of silt accumulation by mechanical means.

Groundwater levels fluctuate according to additions or removals from the aquifer, and precipitation provides the basic means of adding to groundwater storage. During the 1987-1992 drought period, the groundwater supply was depleted each year. However, during the overall period between 1966 and 2000, approximately 1.5 million acre-feet have been added and approximately 0.7 million acre-feet have been extracted, with a net increase in supply of approximately 0.8 million acre-feet.

The effect of the project has been to provide a firm water supply to a large number of irrigators whose well supply was failing or who were pumping groundwater of unsatisfactory quality. In addition, the build-up of the water table has resulted in reduction in subsurface inflow from neighboring areas and a significant improvement in both groundwater depth and water quality for those irrigators in the district who continue to rely on groundwater.

Because of the success of this project, the district has entered into a 25-year agreement with the Metropolitan Water District of Southern California (MWD) whereby the district will bank approximately 250,000 acre-feet of MWD state water project supply and then return the water to MWD in drought years. In order to accomplish this agreement, new facilities are under construction. These facilities include 500 acres of new spreading ground, 15 groundwater wells, and a 4 $^{1}/_{2}$-mile bidirectional inertia pipeline connecting the terminus of the district's south canal with the California Aqueduct. This is an example of a mutual arrangement between the urban and agricultural sectors to stabilize the annual availability of water to offset the impacts of a sustained drought.

Palo Verde Irrigation District-MWD Program

The Palo Verde Irrigation District (PVID) is located adjacent to the Colorado River, north of Blythe, California. Both the PVID and MWD have water rights to the Colorado River. Both the PVID Board of Trustees and MWD Board of Directors have approved the principles for a 35-year program that will pay farmers to annually set aside a portion of their land, rotate their crops, and transfer saved water to urban southern California. The approved principles are based on the successful experience of a two-year program between the two districts during the dry years of 1992-1994. The approved principles specify that Palo Verde valley farmers will stop irrigating from 7 to 29 percent of their land in any year at the request of MWD. A range of 25,000 to 111,000 acre-feet per year of water will be made available to urban customers from the reduction in irrigation. The land taken out of production will be maintained and rotated once every one to three years. Thus every farm will continue to irrigate at least 71 percent of its property and none of the farm's land would be permanently taken out of production. This arrangement is expected to maintain the viability of agriculture and minimize third-party effects.

For each acre set aside, farmers who sign up will receive a one-time payment of $3,170 and then $550 annually beginning in 2002. In addition to payments to individual farmers, MWD has agreed to invest $6 million in local community improvement programs to offset any potential economic impacts from the proposed program.

The primary motivation for this agreement is the USBR requirement that California reduce its use of Colorado River water. Future reductions in Colorado River water availability, accompanied by the projected increase in urban water demand from growing human populations, would be partially accommodated by this agreement.

CONCLUSIONS

The case studies presented from California illustrate the benefits of water storage from wet years to partially offset deficiencies during drought years. They also represent cooperative, mutually beneficial arrangements between the urban and agriculture sectors. The conjunctive use of surface and groundwater supplies is well illustrated in the Arvin-Edison water storage program. Although the urban-agricultural water exchange directs water from agriculture to the urban sector, the farmers receive payment for their water to offset some of the losses that naturally occur during drought periods. Several instruments are available to minimize, but not completely eliminate, the economic losses associated with an extended drought period.

The Economics of Agricultural Water Use and the Role of Prices

David Sunding

Agricultural production depends heavily on the nature of the local environment. Factors such as soil permeability, slope, and microclimate have a large effect on yield and water use. It is important to explicitly incorporate the influence of these "microparameters" when assessing agricultural water demand.

The farmer's choice of irrigation technology can have a large influence on the demand for applied water, so it is sensible to begin with a description of how farmers select irrigation methods. Let

y = output / acre
e = effective input / acre
a = applied input /acre

$$i = 0 \text{ technology indicator } \begin{cases} i = 0 \text{ for traditional} \\ i = 1 \text{ for modern} \end{cases}$$

α = land quality
p = output price
w = input price
k_i = per acre cost of technology $i, k_1 > k_0$.

The crop production function is $y = f(e)$ with $f'(e) > 0$ and $f''(e) < 0$. The input efficiency function, $h_i(\alpha)$, is the fraction of the applied input consumed by the crop under technology on land quality. The technologies are such that

$$0 \le h_0(\alpha) = \alpha \le h_1(\alpha) \le 1$$
$$h_i^{'}(\alpha) > 0 \quad and \quad h_i^{''}(\alpha) < 0.$$

The farmer's irrigation technology choice problem is as follows:

$$\text{(1)} \qquad \begin{aligned} &\max_{\delta_i, a_i} \sum_{i=0}^{1} \delta_i \left[pf\left(h_i(\alpha)a_i\right) - wa_i - k_i \right], \\ &\text{s.t.} \quad \delta_i \in \{0,1\} \\ &\qquad 0 \le \sum_{i=0}^{1} \delta_i \le 1. \end{aligned}$$

The search for a maximum consists of two stages. First, the optimal amount of applied water (a continuous choice) is determined conditional on each technology. Then, working backwards, the highest-profit technology is identified.

The applied input choice is determined by the following:

$$\text{(2)} \qquad \pi_i = \max_{a_i} pf\left(h_i(\alpha)a_i\right) - wa_i - k_i.$$

The Future Operating Capability (FOC) is

$$\text{(3)} \qquad pf' = \frac{w}{h_i}.$$

In words, this optimization condition implies that the Value of the Marginal Product (VMP) of effective water must equal the marginal price of effective water. Once the second-stage, continuous problem is solved, the discrete choice problem of technology selection must be addressed, choosing

$$\begin{aligned} &\delta_1 = 1 && \text{if} \quad \pi_1 > \pi_0 \text{ and } \pi_1 > 0. \\ &\delta_0 = 1 && \text{if} \quad \pi_0 > \pi_1 \text{ and } \pi_0 > 0. \\ &\delta_1 = \delta_0 = 0 && \text{if} \quad \pi_1, \pi_2 < 0. \end{aligned}$$

The model generates a number of testable hypotheses about the influence of environmental and market conditions on adoption of precision technology. Consider first the role of land quality. The marginal impact on profits under technology i of a change in α is as follows:

(4) $$\frac{d\Pi_i}{d\alpha} = pf'h_i'(\alpha)a_i\left[\frac{w\eta_i a_i - a_i}{\alpha}\right] > 0,$$

where $\eta_i = h_i(\alpha)\alpha / h_i(\alpha)$. It follows that the difference in profits between the two technologies is equal to

(5) $$\frac{d\Delta\Pi}{d\alpha} = w\left[\frac{\eta_1 a_1 - a_0}{\alpha}\right].$$

Now, this expression can be signed by taking a Taylor's series approximation of α as follows:

(6) $$a_1 = a_0 + \frac{\partial a}{\partial a}\left(h_1(\alpha) - \alpha\right),$$

recognizing that adoption of the precision technology is equivalent to a shift in land quality from α to $h_1(\alpha)$. Substituting the elasticity expressions above, it follows that

(7) $$a_1 = a_0\left(1 - \frac{1-1/\phi}{\alpha}\right),$$

where $\phi = -f''e / f'$. Substituting this equation into (5), it follows that

(8) $$\frac{d\Delta\Pi}{d\alpha} = w\left[\frac{\eta_i a_0\left(1 - \frac{1-1/\phi}{\alpha}\right) - a_0}{\alpha}\right] < 0.$$

Thus, the profit gap between the modern and traditional technologies decreases as land quality improves. In this sense, the modern technology augments land quality. A further result helps in understanding the influence of land quality on adoption. At the highest possible level of land quality (i.e., $\alpha = 1$), the modern technology will not be adopted. To see this, simply note that at this land quality $h_1(1) = h_0(1) = 1$, and $\Delta\Pi(1) = k_0 - k_1 < 0$.

At some level of land quality, all else being equal, the identity of the highest-profit technology will change. This level of land quality is called the switch point (α^s). Modern technology is adopted for levels of land quality below $\alpha = \alpha^s$ and the traditional technology elsewhere. Note that the modern irrigation technology also has an extensive margin effect in that it enables profitable operation on lower levels of land quality than does the traditional technology (i.e., $\hat{\alpha}_1 < \hat{\alpha}_0$, where $\hat{\alpha}_i = \{\alpha | \Pi_i(\alpha)=0\}$ is the shut-down level of land quality under technology *i*).

Figure 1 shows that it is not profitable to operate on land of quality $\alpha < \hat{\alpha}_i$ regardless of the type of irrigation technology chosen. On the other hand, with high quality land, either technology is profitable, although the traditional technology is more profitable. This is because on high quality land, the increase in yield with the modern technology is not worth the fixed cost of installing it. Where the modern technology makes a difference is on land of moderate quality, i.e., the land between $\hat{\alpha}_i$ and α^s. The modern technology increases profits on land between $\hat{\alpha}_0$ and α^s, and enables production to be profitable on land between $\hat{\alpha}_1$ and $\hat{\alpha}_0$.

With respect to market parameters, total differentiation of the equation implicitly defining the switch point, α^s, reveals that

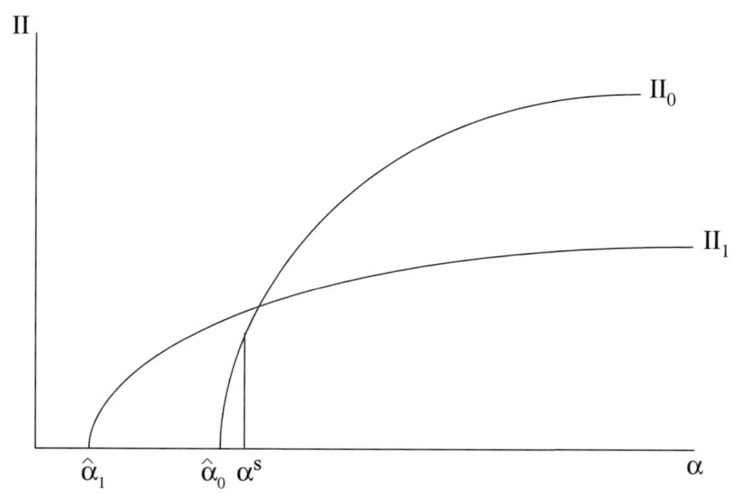

FIGURE 1 Adoption of precision technology.

(9) $$\frac{d\alpha^s}{dW} = \frac{(a_1 - a_0)\alpha^s}{W[\eta_1 a_1 - a_0]} > 0 \text{ if } \phi > 1 \text{ and}$$

(10) $$\frac{d\alpha^s}{dP} = \frac{(y_1 - y_0)\alpha^s}{W[\eta_1 a_1 - a_0]} > 0.$$

OTHER FARM-LEVEL EFFECTS OF PRECISION TECHNOLOGY ADOPTION

Recall that profit maximization requires

(11) $$\frac{w}{h_i} = Pf_e$$

at the optimum. Since $f'(e) > 0$, it follows that

$$h_0 < h_1 \Rightarrow f_0' > f_1' \Rightarrow e_0 < e_1.$$

Thus, modern technology increases the optimal level of effective water use (e). But note that a higher level of effective water use does not imply a higher level of applied water use (a). This is because the ratio of applied to effective water is smaller with modern technology, so that greater effective water can be utilized with lower applied water. In most cases, modern technology reduces the optimal level of applied water use, and is therefore water-saving. If $e_0 < e_1$, then $y_0 < y_1$. Thus, use of the modern technology increases crop output.

If land quality is high, water quality is high, and the weather is mild, then h_1 and h_0 are not very different, and the adoption of modern irrigation technology will have only a small effect on the optimal levels of crop output and applied water.

If land quality is low, water quality is low, or the weather is hot, then adoption of modern irrigation technology may affect optimal crop output and applied water use significantly. When land quality is low and temperature is high, the effect of adopting new technology depends on water price.

EMPIRICAL ANALYSIS OF IRRIGATION TECHNOLOGY CHOICE

Despite the importance placed on micro-level variations in the theoretical literature, most empirical studies of irrigation technology adoption suffer from the use of regional average data on technology choices and resort to comparing percentages of adoption among states or counties. Previous empirical studies have not been able to match technology choice on a one-to-one basis with micro-level variables, such as water-holding capacity, field gradient and size, water price, and water supply source. Averaging data on a regional basis has a homogenizing influence on both grower behavior and physical characteristics; it may obscure the effect of micro-variables, and, as a result, it may seriously bias statistical estimates of adoption behavior.

Empirical Model

The grower decides which irrigation technology to adopt on the jth field by calculating expected profits under each of the i technologies, while taking into account what type of crop is grown and the field's physical characteristics. The grower chooses the technology that maximizes perceived profits, given that crop choice already has been made.[1] In this study, crop and technology choice are modeled as sequential. An alternative assumption would be to model the crop and technology choice simultaneously, as suggested by Negri and Brooks (1990) and by Lichtenberg (1989). While this approach may be appropriate for grain crops, it does not appear to be appropriate for high-value fruits and vegetables. The distinction is that the production of high-value crops involves extremely specialized capital, where grains are not as highly specialized. Therefore, even though the actual investment in a new crop and technology physically may be made at the same time, the decision to invest is made sequentially. To test this, a model of simultaneous crop and technology choice was estimated. The model had inconsistent results, predicted poorly, and was statistically insignificant.

Given the assumption of sequential choice, the per acre profits are given by

$$\pi_{ij} = \beta'_i X_j + \varepsilon_{ij} \tag{12}$$

where β_i is a vector of estimable parameters, X_j is a vector of observed field characteristics (including crop choice), and ε_{ij} is an unobserved scalar associated with unmeasured characteristics. Setting the index of the traditional technology to $i = 0$, the grower selects the i, the modern technology if

[1]Though much of the more general literature on technology adoption examines profit risk, this is not of great concern in the irrigation technology adoption literature. Note that pressurized irrigation technologies generally increase uniformity of input application, decrease output variability, and increase expected yields. The net result of these attributes to risk considerations is ambiguous since they affect risk in opposite directions.

(13) $$\beta'_i X_j - \beta'_0 X_j > \varepsilon_{oj} - \varepsilon_{ij}.$$

To estimate the model parameters, it is necessary to choose a distribution for the ε_{ij}'s and, thus, the distribution of the difference of the error terms. Two common assumptions are either the normal or the Weibull distributions (Domencich and McFadden, 1975). Normal random variables have the property that any linear combination of normal variants is normal. The difference between two Weibull random variables has a logistic distribution, which is similar to the normal, but with larger tails. Thus, the choice is somewhat arbitrary, especially with large sample sizes. We assume that the ε_{ij}'s follow a low Weibull distribution. Given this assumption, the probability that the ith technology is adopted on the *j*th field is

(14) $$P_{ij} = \frac{e^{\beta_i X_j}}{\sum_i e^{\beta_i X_j}}; i = 0, I; \text{ and } j = 1, J.$$

These give the estimation equations for the standard multinominal logit model that is based on the characteristics of the field, not the characteristics of the choice. In this model the parameters vary across technology choices, but not across field characteristics. Thus, the number of estimated parameters is equal to the number of characteristics times the number of choices.

The effect of each of these variables is captured in the estimated parameter vector β. The difference in characteristics across fields affects the technology choice via the perceived effect on the profitability of production on a specific field. This differs from previous studies that have looked at how regional differences affect profitability. While the previous results have given insight to regional differences, they do not correspond to individual grower choices given the field characteristics they face.

Data

The model is applied to the Arvin Edison Water Storage District (the District) located in the southern San Joaquin Valley in central California. Because of the regional climate and favorable soils, growers in the District benefit from an early harvest season that allows for diverse cropping patterns, as shown in Table 1. In addition, there has been a large degree of irrigation technology adoption: 30 percent furrow or flood, 37 percent high-pressure sprinkler, and 33 percent low-pressure drip and micro-sprinkler (Table 1). The distribution of crops and irrigation technologies makes the District ideal for analysis; yet, the area is relatively small, so the growers participate in many of the same markets and institutions.

TABLE 1 Irrigation and Acreage by Crop

		Percentage of Acreage by Irrigation Technology		
Crop	Acreage	Furrow	Sprinkler	Drip
Citrus	2,065	15	1	84
Deciduous	11,700	27	33	40
Grapes	23,665	61	2	37
Truck Crops	27,283	11	86	3
Total	74,713	30	37	33

The data on crop choice, irrigation technology, price of water, and water source were collected by the District. The study considers four crop categories: truck crops, citrus trees, deciduous trees, and grape vineyards. Taken together, these crops constitute 76 percent of the cultivated acreage in the District. The remaining acreage is distributed among grains, irrigated pasture, cotton, and dry land crops.

Irrigation technologies are consolidated into three groups based on the required level of pressurization. These groups are as follows: (1) furrow, flood, and border, which are considered the traditional or gravity technology and are used on all types of crops; (2) high-pressure sprinklers, which are used primarily on truck and deciduous crops; and (3) low-pressure systems like drip, micro-sprinklers, and fan jets, which are also used in each crop group.

There are several important points to be raised concerning low-pressure technologies and perennial crops in the District. First, low-pressure systems such as drip irrigation only wet a small area of soil. As a result, perennial crops under drip irrigation form a smaller root system than if a traditional irrigation system were used. Many growers feel that this makes the crop more susceptible to disease and the accumulation of salts, reducing the attractiveness of these systems. Second, many of the perennial crops were established prior to the introduction of low-pressure systems. Because different types of root systems are developed under the different types of technologies, growers are reluctant to switch technologies on an established crop for fear of damaging the crop. To combat these potential problems, growers have used multiple emitters for each tree to achieve a larger area of water dispersion.

The marginal price of groundwater is estimated by the District based on depth to groundwater and the energy cost for the size of pump needed to lift water from a given depth. The marginal price for surface water is the variable component of the District charge for each acre-foot that is actually delivered. In 1993, marginal water price ranged from \$2 to \$57 per acre-foot for surface water, and \$40 to \$88 per acre-foot for groundwater. Though the marginal price of groundwater is about \$25 more per acre-foot than surface water, the fixed component of the District charge for surface water is set so that the total price for ground and surface water is approximately the same, ranging from \$50 to \$110 per acre-foot.

The Kern County Natural Resource Conservation Service collected data on soil permeability and field slope to define land quality for each quarter section. To match the quarter sections (which are 160-acre plots) to the specific fields, District land maps were used to identify the exact location of each field. Permeability and slope were given in inches per hour and percentage, respectively. The data indicate that the distribution of irrigation technology varies by slope; when the slope increases so does the percentage of acreage under drip irrigation. This indicates that the grower's irrigation technology choice is conditioned on land characteristics. The effect of soil permeability on technology choice is not as distinct.

Estimation

The econometric model explains the use of the different types of irrigation technologies as a function of the characteristics of the fields for which they are used. The estimation equations in (14) provide a set of probabilities for the $I + 1$ choices faced by the decision maker. However, to proceed it is necessary to remove an indeterminacy in the model. A convenient normalization is to assume that β_0 is a vector of zeros. We can then take the log and estimate the log odds ratio of choosing the *i*th technology on the *j*th field. This is given by

$$\ln\frac{P_{ij}}{P_{0j}} = \beta'_i X_j,\ i = 1,2,\ \text{and}\ j = 1,2,\ldots,1,493. \tag{15}$$

The coefficients can be interpreted as the marginal impact of the variable on the log odds of selecting a modern technology relative to the benchmark (traditional) technology.

The data for the study are from the 1993 growing year, and there are 1,493 fields cultivated by approximately 350 growers. Though we are unable to identify which growers cultivated which fields, based on sample interviews we determined that most growers had fewer than four fields and grew at least two different crops. Growers that had a large number of fields grew at least five crops. There are eight independent variables, four continuous, and four binary. The four continuous variables are (a) field size, (b) field slope, (c) soil permeability, and (d) price of water. The four binary variables are (e) water source (i.e., groundwater or both ground and surface water), (f) citrus crop, (g) deciduous crop, and (h) grape vineyard. Without loss of generality, truck crops and gravitational technology are used as benchmarks for crops and technology choice.

Estimation Results

The Limdep statistical package is used to estimate the parameters of the model using maximum likelihood estimation and Newton's method. We report the coefficients, asymptotic t-statistics, and three statistical tests to evaluate the performance of the model. To allow comparison of adoption rates among traditional, sprinkler, and drip technologies, we calculate the probability of adoption, the elasticity of the continuous variables, and the percent change in probability of the discrete variables if they were to change from 0 to 1. These are all reported in Table 2.

Of the coefficient estimates in Table 2, more than half were significant at the 0.0001 level, and all but two were significant at the 0.07 level. To measure the performance of the model, the McFadden R^2, the log-likelihood ratio test, and the percentage of correct predictions are reported. The McFadden R^2 is calculated as $R^2 = 1 - L_\Omega / L_\omega$, where L_Ω is the restricted maximum log-likelihood and L_ω is the restricted maximum log-likelihood with all slope coefficients set equal to zero (Amemiya, 1981). The log-likelihood ratio test is given by $2(L_\Omega - L_\omega)$ and is asymptotically distributed as a chi-squared random variable. The percentage of correct predictions is calculated as the total number of correct predictions as a percentage of the number of observations. Each of these measures indicated that the model has strong explanatory power.

The statistical results indicate that the adoption of irrigation technologies is highly dependent on crop choice. The coefficients on the perennial crop variables in the sprinkler technology equation are all negative, large, and highly significant. This result implies that the probability of adopting sprinkler rather than the traditional technology is low for perennials and reflects the physical characteristics of perennial crops. For example, high-pressure sprinklers disperse water over a large area, saturating the tree and causing fruit decay, which is not a problem for many annual crops such as potatoes. Crop choice also strongly affects drip adoption, although in nearly the opposite way as for sprinklers. Perennial crops, especially citrus trees, are more likely to be grown under drip irrigation than annuals. The influence of crop type on technology choice is also reflected in the change in probability figures in Table 2. These results show that a grower producing perennial crops is much more likely to adopt drip than furrow or sprinkler irrigation. For example, growing citrus trees increases the probability of adopting drip by 58 percent, holding all other variables at their mean value. Previous studies that focused on a small number of crops (Lichtenberg, 1989; Shrestha and Gopalakrishnan, 1993) could not fully identify the importance of crop type on irrigation technology adoption.

Economic factors are also important in determining irrigation technology choices. The coefficient on the water price variable in the drip equation is positive and significant, confirming previous findings that water-saving technology will be adopted as water price increases. However, the coefficient on water price

TABLE 2 Estimation Results, Elasticities and Probabilities

Variable	Estimation Results[a]		Elasticities[b]		
	Sprinkler	Drip	Furrow	Sprinkler	Drip
Constant	1.9855	–4.5480			
	(3.372)	(–7.701)			
Water Price ($/acre-foot)	–0.0130	0.0257	–0.24	–0.84	0.96
	(–1.333)	(3.151)			
Surface water (0/1)	–0.5099	0.9706	[–0.11]	[–0.12]	[0.23]
	(–1.636)	(3930)			
Soil permeability (in/hr)	0.0002	0.0529	–0.04	–0.04	0.11
Field slope (%)	0.2210	0.6277	–0.32	0.01	0.61
Field size (acres)	0.0101	0.0065	–0.19	0.34	0.15
Crops					
Citrus (0/1)	–5.1537	2.1117	[–0.21]	[–0.37]	[0.58]
	(–8.380)	(6.095)			
Deciduous (0/1)	–2.3600	1.3872	[–0.16]	[–0.23]	[0.39]
	(–11.186)	(4.064)			
Grapes (0/1)	-6.3777	0.6760	[0.24]	[-0.57]	[0.33]
Probability of adoption evaluated at variable means			0.54	0.18	0.28
Observations	1,493				
McFadden R^2	0.44				
Likelihood ratio test x_{16}^2	1,441.16				
Correct prediction	74%				

[a]Terms in parenthesis are asymptomatic t-statistics.
[b]Terms in brackets are not elasticities. They are the percent change in the probability of adoption as the discrete variable changes from 0 to 1.

in the sprinkler equation is negative. Figure 2 shows the change in the probability of adoption as a function of the price of water, with all other variables set at their mean values. This figure demonstrates that, as the price of water increases, growers switch from both furrow and sprinkler irrigation technologies to drip.

The results in Table 2 and Figure 2 are in sharp contrast to the results of previous studies that have found similar adoption patterns for high- and low-pressure irrigation systems. For example, Caswell and Zilberman (1985) reported coefficients of 0.03 on marginal water price in equations explaining both drip and sprinkler adoption, and Cason and Uhlaner (1991) estimated water price coefficients between 0.02 and 0.07 for all technologies, depending on the region. The results differ from these studies for several reasons. Examining several technology choices simultaneously gives a more complete picture of grower decision-making behavior and allows for explicit estimation of marginal probabilities. Further, growers in this study farm in an arid, hot climate and pay more for water than irrigators in many other areas. As a result, the diffusion process for pressurized technologies is more advanced in the District than in other regions, and sprinkler technologies appear to be nearing the end of their product life cycle. Sprinkler irrigation has been employed in the District since the early 1960s and is widely utilized on crops that grow well with this technology. In particular, Table 1

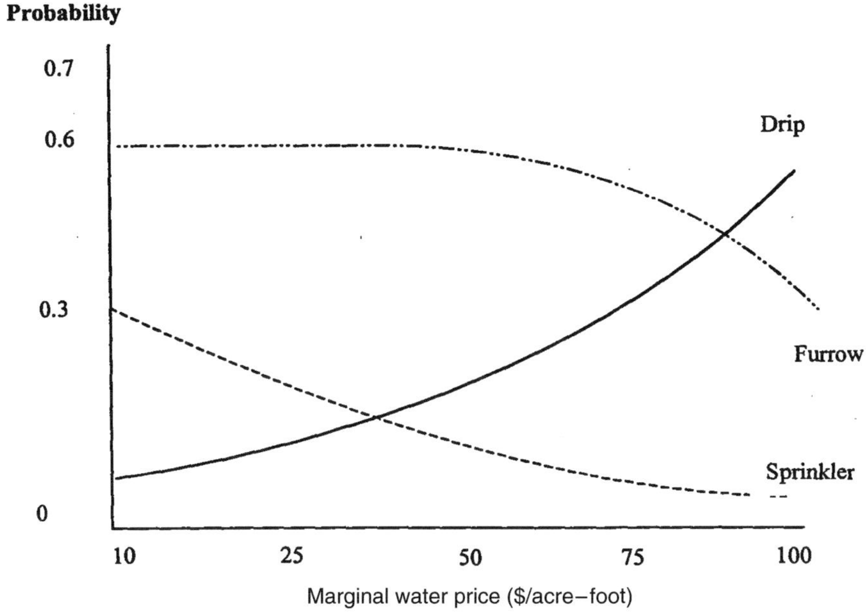

FIGURE 2 Probability of adoption by marginal water price.

shows that truck crops are grown largely under sprinkler irrigation. However, potato growers in the District are now beginning to convert to low-pressure systems (especially drip tape) in response to changes in water price. This observation is consistent with the findings of Dinar and Yaron (1992). In their model of technology adoption and abandonment, Dinar and Yaron estimate the technology cycle of hand-move sprinklers to range from 22 to 24 years.

The coefficients on the land quality variables, soil permeability and field slope, are of the expected sign and magnitude. Again, however, there are important differences between technologies in terms of the effect of land quality variables. Sprinkler adoption is not as sensitive to land quality as drip irrigation, which is especially dependent on field slope. Prior to the introduction of drip irrigation, it was difficult and costly to grow irrigated crops on lands with steep slopes. As a result, the introduction of drip has allowed cultivation of land that had previously been unproductive. This relationship is best seen in Figure 3, which shows that variations in slope have a dramatic effect on the probability of adopting furrow and drip irrigation.

Caswell and Zilberman (1986) show theoretically that modern irrigation

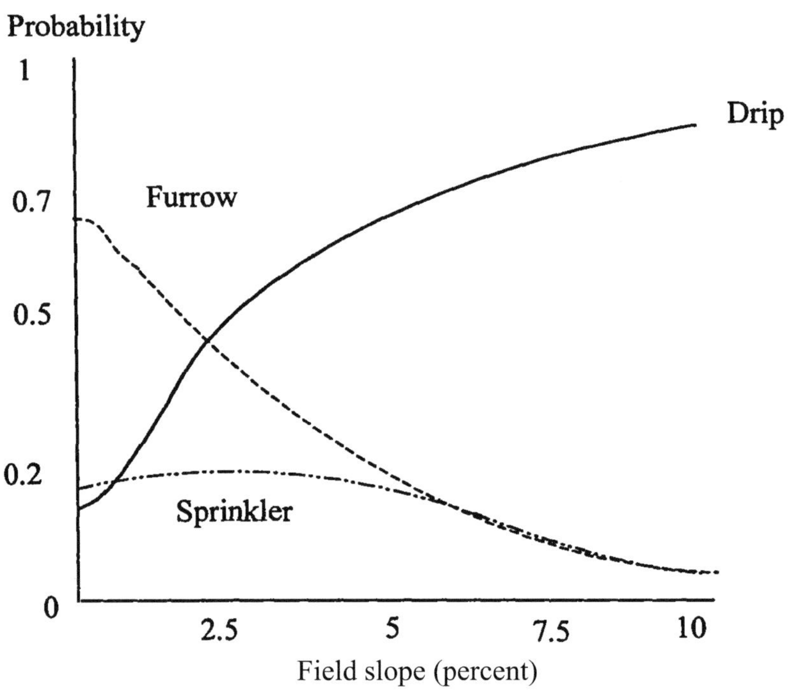

FIGURE 3 Probability of adoption by field slope.

technologies are less likely to be adopted on fields with surface water supplies rather than groundwater supplies on the assumption that surface water is supplied at lower pressure than groundwater. The statistical results show that sprinkler adoption is less likely in areas with surface water supplies, but that drip adoption is more likely with surface supplies. While the District is one of the few California districts supplying pressurized surface water to its growers, the pressure is not consistent and is only sufficient to run a low-pressure system such as drip.

Discussion and Implications

These empirical results point out that cross-section technology adoption coefficients must be interpreted with the dynamic diffusion process in mind and also show that the effect of economic factors such as price on adoption is path-dependent. For example, in the results, we obtained a negative coefficient on the water price variable for adoption of sprinkler irrigation, which would seem to refute the theoretical and empirical literature. However, high-pressure sprinklers are widely adopted in the study area, and because these technologies are far from the beginning of their life cycle in the District, abandonment of sprinkler technologies is more sensitive to water price increases than is adoption. In another area where growers rely more on gravitational systems, and hence sprinklers are at the beginning of their life cycle, the opposite should be true. This demonstrates that the coefficients cannot be interpreted at face value and that it is important to consider the underlying diffusion process when considering the policy implications of an analysis.

The results show that water price is not the most important factor governing irrigation technology adoption; physical and agronomic characteristics appear to matter more. As a result, the distributional impacts of irrigation water pricing reforms will be significant, with changes in producer welfare following the spatial distribution of environmental characteristics. To the extent that micro-level factors condition irrigation technology choice, policies that change the price of irrigation water to reflect its off-farm value will result in a pure loss for some producers while encouraging adoption of modern irrigation technologies for other producers. This demonstrates the importance for economists to bear in mind the equity implications of water pricing reform proposals when interacting with decision makers.

This type of empirical research has important implications for the design of water pricing and delivery policies. The statistical results above show that large increases in the price of water generally stimulate the adoption of drip irrigation systems; that adoption patterns are heavily influenced by crop type; and that the adoption decision is also strongly conditioned by slope, but is only slightly affected by variations in water-holding capacity. These results are a significant departure from previous studies, which have generally failed to account for differences in adoption behavior within the group of pressurized technologies and for the influence of crop type on adoption behavior, and which have inadequately measured physical characteristics and water prices by relying on regional data.

WATER ALLOCATION MECHANISMS AND AGRICULTURAL WATER USE

In many regions of the United States, water allocation has been based on queuing systems rather than on markets. Queuing systems are sets of laws defining property rights regarding who has priority to use water, when water may be used, how water may be used, and how much water may be used. Although queuing systems are still the norm in many parts of the world, they are undergoing change. A typical queuing system is a use-it-or-lose-it system of water property rights based on the principle "first come, first served."

Queuing systems for water were established in the western United States to encourage settlement of land and the economic use of water resources. In early periods, water was abundant, governments were poor, and they wanted to encourage people to develop water resources. As a result, government gave individuals the right to the water that they diverted, so long as they used it in a way that generated economic benefits. Note that markets are the best allocation mechanisms when there is scarcity, but queuing can be very effective when scarcity does not exist. The biggest problem with queuing is that demands for water can increase, and when that happens, scarcity suddenly can emerge. In this situation, water reforms are needed.

As water scarcity increases there is a tendency to allow trading in water. In California, for example, water trading was introduced during drought periods. The transition from queuing to markets may involve redesign of the water allocation system, building a system for the monitoring of water use, and protection against theft, and all this entails high transaction costs. If the gains from transition are smaller than these transaction costs, reform will not and should not occur.

The queuing system is not an efficient means of allocating water resources if scarcity exists, that is, if junior rights holders do not receive enough water. In particular, if a unit of water provided to junior rights holders has positive Marginal Value of Product (MVP), then we know queuing is inefficient since senior rights holders apply water to the point where their MVP = 0. Figure 4 illustrates this argument.

Because the prices facing senior and junior users are unequal, the corresponding marginal benefits of water use are unequal. Since the marginal cost to supply each type of user is essentially the same, social welfare may be increased by reallocating water from senior users to junior users. Thus, allocation by queuing is inefficient. Allowing water to be freely traded would lead to water transfers from senior rights holders to junior rights holders.

Notice also that the total level of water consumed is inefficiently high under

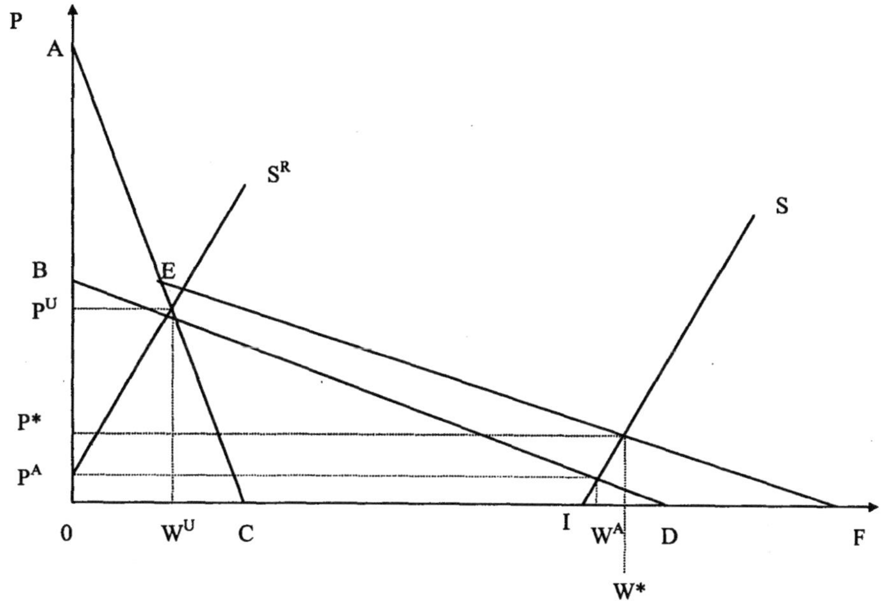

FIGURE 4 Queuing versus markets.

(a) Water supply projects (dams, canals, etc.) often have high initial fixed costs associated with construction and low marginal costs of supplying water up to the capacity of the project. At full capacity, the marginal cost of water supply rises steeply, because additional projects or procurement strategies are required in order to supply additional water. Thus, the marginal cost of water supply curve is 0IS.

(b) Assume that senior water demand is given by curve BD.

(c) Assume that junior water demand is given by curve AC.

(d) Aggregate demand for water is given by curve AEF, if water markets exist. Under water markets, the equilibrium level of water consumed is W* and the equilibrium price is P*.

(e) Water rights allocate water to different users at different times. Demand is not aggregated, but discriminated by time in the residual demand curves BD and AC. Senior rights holders purchase an amount of water equal to W^A, which is where their water demand equals the marginal cost of water. The price of water used by senior rights holders is P^A.

(f) Once the W^A units of water have been consumed by senior users, junior users face residual water supply S^R, and therefore consume W^U units of water and pay a price of P^U. The price of water in junior rights areas is higher than in senior rights regions.

a queuing system, or that $W^A + W^U > W^*$. Thus, moving to a market oriented system of water allocation can lead to greater water conservation.

Queuing System

We now return to the analysis in the first section of this paper and demonstrate some economic implications of the transition from queuing to markets. Water trading is disallowed under a queuing system. In this case, there is no incentive for senior rights holders to adopt modern technology since water has no price. Water is simply diverted, as needed, according to a farmer's place in the queue.

Suppose that A is the total amount of water available in a region and that L is the total amount of land available for cultivation. Also, let $y_m = f(e_m)$ denote maximum output per acre, where e_m is defined by $e = \{e|f'(e) = 0\}$ is the effective water associated with maximum yield.

Senior rights owners use water until the Value of the Marginal product (VMP) of water = 0, which is the level that will maximize yields. Settlement occurs until water resources are exhausted, so water becomes the limiting factor on development. Applied water use is $a_m = e_m / h_0$ per acre, the amount of applied water associated with the maximum effective water absorbed by the crop.

Under a queuing system of water rights, the water price is equal to 0, and the per-acre fee for water use μ. Settlement will occur until all water is appropriated and the total acres under cultivation is $A / a_m = Ah_0 / e_m < L$. Total output under a water rights system is $Ah_o y_m / e_m$ and producer surplus is $PAh_0 y_m / e_m - \mu Ah_0 / e_m$.

Market System

When all land quality is the same, the efficient solution involves applying water uniformly across all land to equate the MVP. Thus, under a market system, all land is utilized, and each owner faces the choice of technology *i*.

Under a market system, water use per acre is *A/L* yield per acre under technology *i* is $y_i = f(h_i A / L)$, and the price of water VMP of applied water = $Pf_e h_i$.

Producers' annual profits per acre are

$$\Pi_i = Py_i - \frac{A}{L} Pf_e h_i - k_i - \mu - t \qquad (16)$$

where *t* equals transaction costs so that

(17) $$\Pi_1 - \Pi_0 = P(y_1 - y_0) - \frac{A}{L} Pf_e(h_1 - h_0) - (k_1 - k_0).$$

Technology 1 is selected if $\Pi_1 - \Pi_0 > 0$.

Both technologies require the same water per acre, because water is evenly distributed across all acres as a result of equating the MVP. When each farmer is a small unit, the farmer does not believe that her choice of technology will affect the market price of water, and the market price of water is considered as a constant in the problem. Then the choice of technology can be expressed as follows:

Select technology 1 when $p(y_1 - y_0) > k_1 - k_0$.

Both technologies result in the same water use per acre, but the modern technology increases the yield by raising the amount of effective water received by the crop. If the market value of the increase in yield is greater than the extra capital costs involved with investing in the new technology, then the farmer should invest.

COMPARING MARKET AND QUEUING OUTCOMES

Assume that, under market conditions, technology i is optimal and adopted by all farmers. Under a market system all arable land is utilized. The transition to market will increase irrigated land from Ah_0 / e_m to L. Output will increase by

(18) $$\Delta Y = Lf\left(h_i \frac{A}{L}\right) - \frac{Ah_0}{e_m} f(e_m) > 0$$

as water is shifted from low Marginal Product (MP) land to the high MP lands now under cultivation. Output per acre will decrease from $f(e_m)$ to $f(A / Le_i)$. The reduction in water use per acre is from $e_m \cdot h_0$ to A / L.

In the transition to the market, $(e_m - h_i A / L) h_0 A / e_m$ units of water that were used to produce the output associated with area B in Figure 5 are allocated to irrigate new lands. Overall, output increases because the water that was used under queuing to produce output associated with region B of Figure 5 is used under markets to produce output in region A on new land that is brought into production. Obviously, the marginal productivity of this water increases.

If the senior rights owners who appropriated water under the queuing system have to buy water after the transition to a market, their profits will decrease. Under water markets, they now have lower yields, they now have to pay for

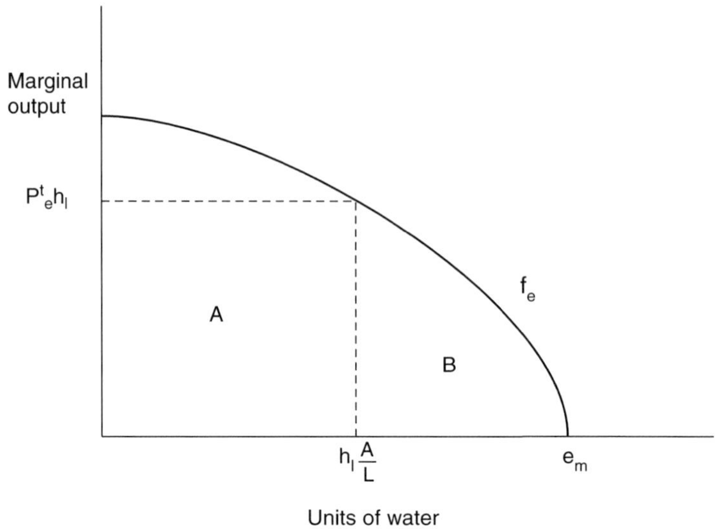

FIGURE 5 Output effect of transition to markets.

water, and they also must pay to adopt the new technology, since doing so is now optimal. Their loss per acre is

$$Pf(e_m) - Pf\left(h_i \frac{A}{L}\right) + Pf_e h_i \frac{A}{L} + k_i + \mu. \tag{19}$$

But if the senior rights holders are given the property rights to the water, they may gain. These users still have lower output than under queuing, but the gain from selling water may overcome this output loss. Their income per acre will be

$$Pf\left(h_i \frac{A}{L}\right) + \left(e_m - h_i \frac{A}{L}\right) Pf_e h_i \frac{A}{L} - t - k_i. \tag{20}$$

If the transaction costs are high, there may be no reason to switch to a system of water markets. Namely, if

$$t > \frac{P\Delta Y + k_i L - k_0 A / (h_0 \cdot e_m)}{L}, \tag{21}$$

then transaction costs per acre exceed the per acre change in output plus the cost of adopting the optimal modern technology less the cost savings of senior owners not adopting the conventional technology. In this case, water markets are inefficient.

Because markets for final products have negatively sloped demand, the transition from queuing to markets will also reduce the market price of agricultural commodities. Senior rights owners may thus lose, even if they sell water because of the price decline of their output. Producers as a whole may actually lose, but consumer surplus will increase.

WATER PRICING IN A CONJUNCTIVE USE SYSTEM

The economics of agricultural water use and pricing become more complicated when the possibility of groundwater use is considered. Let y_t = pumping in year t and x_t = the level of the groundwater stock in year t. The level of the stock increases in year t by $g(y_t)$, where $\partial g / \partial y_t \geq 0$. For example, $g(y_t) = \Lambda + \Theta y$, where A is rainfall plus imported surface water (which is influenced by the price of water charged by the government) and Θ is percent of irrigation water that is return flow. Thus, the growth equation for the stock of groundwater is

(22) $$x_{t+1} = x_t + g(y_t) - y_t.$$

Let $B(y_t) =$ benefits from groundwater use and $C(x_{t1} y_t) =$ total cost of pumping, where $\partial c / \partial x_t = 0$ and $\partial c / \partial y_t \geq 0$.

Groundwater is managed optimally when pumping and the stock are chosen each year to maximize

(23) $$\sum_{t=0}^{\infty} (1+r)^{-t} \left[B(y_t) - C(x_{t,} y_t) \right] \quad t = 0,1,2...$$

where r is the price of land, subject to $x_{t+1} = x + g(y_t) - y_t$. To solve this problem, convert the constrained problem into an unconstrained one using Lagrange multipliers as follows:

(24) $$\max_{x_t, y_t, \mu_t} \sum_{t=0}^{\infty} (1+r)^{-t} \left[B(y_t) - C(x_t, y_t) \right] + \mu_t \left[x_t + g(y_t) - y_t - x_{t+1} \right].$$

The first order conditions are

(25) $$\frac{\partial L}{\partial y_t} = (1+r)^{-t}\left[\frac{dB}{dy_t} - \frac{\partial c}{\partial y_t}\right] + \mu_t\left[\frac{ds}{dy_t} - 1\right] = 0$$

where *s* is units of surface water

(26) $$\frac{\partial L}{\partial x_t} = (1+r)^{-t}\frac{\partial c}{\partial x_t} + (\mu_t - \mu_{t-1}) = 0$$

(27) $$\frac{\partial L}{\partial M_t} = x_t + g(y_t) - y_t - x_t - x_{t+1} = 0$$

(28) $$\text{Define } \lambda_t = (1+r)^t \mu_t$$

λ_t is the value in period *t* of an extra unit of groundwater stock in period *t*. μ_t is the value in period 0 of an extra unit of groundwater in period *t*.

We can rewrite (25) as

(29) $$\frac{dB}{dy_t} - \frac{\partial c}{\partial y_t} - \lambda_t\left[1 - \frac{dg}{dy_t}\right] = 0$$

So optimal pumping equates the marginal benefit of pumping with its marginal cost plus user cost of pumping adjusted for return flows to the aquifer.

Now, from (28) it follows that

(30) $$\mu_{t-1} = (1+r)^{1-t}\lambda_{t-1} = (1+r)^{-t}(1+r)\lambda_{t-1}.$$

Substituting into (26) we get

(31) $$(1+r)^{-t}\lambda_t - (1+r)^{-t}(1+r)\lambda_{t-1} - (1+r)^{-t}\frac{\partial c}{\partial x_t} = 0.$$

Multiplying by $(1+r)^{-t}$ and rearranging yields

(32) $$r\lambda_t = (\lambda_t - \lambda_{t-1}) - \frac{\partial c}{\partial x_t} = 0.$$

This condition says that the optimal stock of groundwater occurs when the opportunity cost of keeping a unit of water in the ground ($r\lambda_t$) equals the capital gain when water is kept in the ground ($\lambda_t - \lambda_{t-1}$) minus the reduction in pumping cost from having more water in the ground.

In a steady-state,

(33, 34) $$x_t = x_{t+1} = x^* \\ \lambda_t = \lambda_{t+1} = \lambda^*.$$

In this case, optimality conditions (22), (29), and (32) reduce to the following:

(22′) $$y^* = g(y^*)$$

(29′) $$\frac{dB}{dy} - \frac{\partial c}{dy} - \lambda^* \left[1 - \frac{dg}{dy}\right] = 0$$

(32′) $$\lambda^* = -\frac{\partial c / \partial x}{r}.$$

I will consider the effect of changes in surface water prices in the context of a steady-state.

When $g(y_t) = A + \Theta y$, and is influenced by the price of surface water, it is clear first of all that raising the price of surface water reduces A, reduces the amount of groundwater pumping, and reduces total water use. The shadow price of groundwater, λ^*, decreases (from equation (32′), and the stock of groundwater decreases (from (29′)). All of these effects must be considered when evaluating the optimality of water price reforms in a conjunctive use setting.

AGRICULTURE AS A SUPPLIER OF LAST RESORT: AN EXAMPLE FROM CALIFORNIA

Agriculture in the western United States is highly dependent on the diversion of water resources for irrigation. At the same time, population growth, increased industrialization and, most importantly, heightened public awareness of environmental benefits from enhancing instream flows are all exerting tremendous pressure on federal and state agencies to reduce these diversions. These forces are resulting in a large-scale reallocation of water from agriculture to urban and environmental uses. What is the effect of this reallocation on agriculture?

A Conceptual Model of the Economic Impacts of Water Supply Reduction

The modeling framework applied here is drawn from Sunding et al. (2001), and consists of a microeconomic model of resource allocation by the irrigated agricultural sector. Profit maximization in agriculture is conducted subject to water supply reductions and economic relationships.

Following on the discussion above, it is crucial to recognize heterogeneity among producers. In this conceptual framework, farming is carried out by J micro production units of various sizes. Such units may be interpreted as farms, water districts, or counties depending on the application and the data available. The micro production unit indicator is j, $j = J$; and the land base of each unit is denoted by L_j. It is assumed that there are no constraints on water movement within the micro production units, but there may be barriers to trade and transfer of water between micro production units. Indeed, water rights regimes, such as the prior appropriation system and riparian rights systems, restrict trading; one major feature of a policy reform is the extent to which water trading is allowed.

The analysis is conducted for N + I water policy scenarios, with n as a scenario indicator $n = 0, 1, 2, ..., N$. The scenario $n = 0$ corresponds to the preregulation or base water allocation. Under each scenario, micro production units are aggregated into regions. Water trading is feasible within regions but not between regions. Let K^n be the number of regions under scenario n, and k^n be the region indicator, so that $k^n = 1, ..., K^n$. The set of micro production units in region k^n is denoted by R_k^n. For example, if we have eight micro production units divided into two regions under scenario n,

$$R_1^n = \{1,2,3,4\},\ R_2^n = \{5,6,7,8\}.$$

Each micro production unit has an initial "endowment" of surface or groundwater representing annual surface water rights and groundwater pumping capacity. Let $\overline{S}_j$ be annual surface water available to district j in the base scenario and $\overline{G}_j$ be annual groundwater available to district j. Alternative policy scenarios affect these water availability constraints.

In the base scenario, total water available to region k^n is

$$\sum_{j \in R_k^0} \left(\overline{S}_j + \overline{G}_j \right)$$

However, surface water availability differs among alternative scenarios. Let ΔS_k^n be the reduction of water supply available to region k^n. The overall surface water supply reduction in scenario *n* is

(35)
$$\Delta S^n = \sum_{k^n=1}^{K^n} \Delta S_k^n.$$

This change reflects the total amount of water reallocated from agriculture. Actual use levels of ground and surface water at region *j* are denoted by G_j and S_j, respectively, with $S_j < \overline{S_j}$ and $G_j < \overline{G_j}$.

Following theory and empirical evidence, Sunding et al. (2001) suggest that California growers have responded to reductions in water supply by (i) changing land allocation among crops (including fallowing), (ii) increasing the amount of groundwater pumping, and (iii) modernizing their water application methods (on this point, see also Moreno and Sunding, 2001; Green and Sunding, 1997; Green et al., 1996; and Zilberman et al., 1995). The modeling of production relationships in Sunding et al. (2001) considers all of these possibilities. There are *I* crops and *i* is the crop indicator, $i = 1, I$. Let the amount of water applied to crop *i* in micro production unit *j* be denoted by A_{ij}, and let L_{ij} be the amount of land allocated to the production of crop *i* at micro production unit *j*. Let Y_{ij} be the output of crop *i* at micro production unit *j*. For modeling convenience, total output is represented as the product of yield per acre, y_{ij}, and acreage of crop *i* in micro production unit *j* is $Y_{ij} = y_{ij} L_{ij}$.

Output is produced by land, labor, irrigation equipment, and other inputs (e.g., chemicals), and is affected by local environmental conditions. The general specification of the per acre production function is

(36)
$$y_{ij} = f(L_{ij}, a_{ij}, z_{ij}, \theta_{ij}),$$

where

$a_{ij} = A_{ij} / L_{ij}$ (applied water per acre),

$z_{ij} = Z_{ij} / L_{ij}$ (annual irrigation equipment cost per acre),

Z_{ij} = total irrigation equipment cost on crop *i* in micro production unit *j*,

and

θ_{ij} = regional environmental quality parameters.

This specification is consistent with the observations of Dinar and Zilberman (1991). Specifically, they argue that increased annual irrigation equipment costs increase output by increasing irrigation efficiency, and that both land quality (in particular, water-holding capacity) and water quality (especially salinity) affect the productivity of water. Specific applications may have special functional forms, but all specifications maintain concavity. Yield per acre may decline as land use increases (i.e., $\partial y_{ij} / \partial L_{ij} \leq 0$) because of decreasing marginal productivity of land.

Let the cost of surface water at micro production unit *j* be W_j^s and cost of groundwater be W_j^g.[2] Generally, $W_j^g > W_j^s$, so that surface water is cheaper than groundwater. The cost of inputs other than water and irrigation technology is assumed to be a convex function of crop *i* acreage in micro production unit *j* and is denoted by the function $C_{ij}(L_{ij})$ with

(37) $$\frac{\partial^2 C_{ij}}{\partial L_{ij}^2} \geq 0.$$

This cost function reflects the important empirical observation that land fertility is heterogeneous in California and that increases in acreage lead to increased expenditures on inputs, such as fertilizers, that augment land productivity.[3]

The most general specification of output markets would assume that producers face downward sloping demand curves and that output prices are determined endogenously. In this case, the optimization problem will maximize the sum of producer and consumer surplus subject to resource constraints. In our model, we assume price-taking behavior and denote the price of output *i* by P_i. This assumption is consistent with the high demand elasticity that California producers face.

Assuming profit-maximizing behavior by growers, the aggregate regional optimization problems under scenario *n* are

(38) $$\Pi^n = \max \sum_{j=1}^{J} \sum_{i=1}^{I} P_i Y_{ij} - W_j^s S_j - W_j^g G_j - Z_{ij} - C(L_{ij}),$$

(39) $$\text{s.t.} \sum_{I=1}^{I} A_{ij} = S_j + G_j \qquad \forall j,$$

[2]These costs are delivery costs or water costs paid by users. Since we are interested in developing a regional organizational model that will provide competitive outcomes, we do not consider differences between private and public costs of obtaining water.

[3]We distinguish between dimensions of land quality, such as water-holding capacity, that affect productivity indirectly (for example, through their effect on the productivity of applied water) and other dimensions, such as fertility, that affect productivity directly.

(40) $$\sum_{j\in R_k^n}\left(S_j + G_j\right) \le \sum_{j\in R_k^n} \overline{S_j} + \overline{G}_j - \Delta S_k^n \qquad \forall k,$$

(41) $$\sum_{i=1}^{I} L_{ij} \le L_{ij} \qquad \forall j.$$

Constraint (39) states that total water used in crops is composed of either surface water or groundwater. Condition (40) is the most important constraint, as it sets a limit on the water available to each region under a given policy scenario. Availability is the sum of water available to districts under initial allocation minus the amount diverted under the specific scenario. Inequality (41) is the land availability constraint.

The solution of the regional optimization problem using Kuhn-Tucker conditions requires assigning shadow prices for each of the constraints. The shadow price of equation (39) is W_j^d. This is the shadow cost of water delivery and is equal to W_j^s if only surface water is used and W_j^g if groundwater is used in district *j*. The shadow price of the regional water constraint (40) is V_k^n. Thus, the marginal cost of a unit of water in district *j* that belongs to region *k* under scenario *n* is $W_j^d + V_k^n$.

If the production function is differentiable, optimal water use per acre with crop *i* at district *j* is at the level where the value of marginal product of water is equal to the shadow price of water.

(42) $$P_i \frac{\partial f_{ij}}{\partial a_{ij}} = W_j^d + V_k^n \qquad \forall_{i,j}.$$

Optimal irrigation cost per acre is determined similarly at the level where the value of marginal product of the expenditure is equal to its price. The next condition is

(43) $$P_i \frac{\partial f_{ij}}{\partial z_{ij}} = 1.$$

The shadow price of the land availability constraint in district *j* is r_j, and under standard assumptions, land is allocated to crop *i* in district *j* so that the value of marginal product of land is equal to r_j, i.e.,

(44) $$r_j = P_i f_{ij}(\ldots) - \left(W_j^d + V_k^n\right) a_{ij} - z_{ij} - \frac{\partial C_{ij}}{\partial L_{ij}} + P_i L_{ij} \cdot \frac{\partial f_{ij}}{\partial L_{ij}} \forall i, j \qquad L_{ij} > 0.$$

Condition (44) states that the optimal acreage of crop i at district j is such that net marginal benefit of land is equal to its shadow price. Marginal net benefits of land are the difference between revenue added by marginal land and the extra cost of water, irrigation technology, and other inputs, as well as the extra cost associated with the decline of land productivity. The conditions are more elaborate if there are land availability constraints for individual crops.

In principle, conceptual and empirical analysis requires solving the model under scenario 0, the initial condition, and then under each alternative scenario. The net income effect of a policy under the scenario denoted by $\Delta\Pi^n$ is the change in producer surplus between scenario 0 and scenario n, i.e.,

(45) $$\Delta\Pi^n = \Pi^0 - \Pi^n.$$

It is expected that, for most scenarios, $\Delta\Pi^n$, namely, reduction in water supply, reduces overall income. But different scenarios assume different partitions of the regions. Under the initial scenario $(n = 0)$, the state is divided into K_0 regions, where water trading is feasible within regions and where water trading is allowed between regions. Two types of scenarios are likely to be associated with a given reduction in overall surface water supply. Under water trade scenarios, trading is allowed throughout the state; under proportional cuts scenarios, the supply reductions to regions are proportional to initial allocations so that the reduction in surface water for regions under such scenarios, ΔS_k^n, is

(46) $$\Delta S_k^n = \Delta\bar{S}\frac{S_k^0}{\sum_{k=1}^{K_0} S_k^0}.$$

By the La Chatelier Principle, given total supply reduction, aggregate profit is higher under the free trade scenario, as there are fewer constraints. In some cases, a water reform that reduces surface water supply and allows trading may increase profit $\left(\Delta\Pi^n > 0\right)$ if gains from trading are greater than losses from surface water supply reductions.

Standard welfare analysis considers impacts on consumer and producer surplus, but policy makers may be interested in changes in other variables.[4] Other such variables are gross farm income, regional income, and employment.

The gross income effect of scenario $n, \Delta R^n$, is derived by subtracting gross revenues of scenario n from gross revenues of the initial scenario. As with net

---

[4]These impact measures were requested from us by the U.S. EPA for their use in designing water quality standards.

income, it seems that gross revenues will decline as aggregate water levels decline. However, under some scenarios, the reduced water supply may lead producers to adopt modern irrigation technologies, which tend to increase per acre yields (Caswell and Zilberman, 1986) but also entail higher production costs. Under these scenarios, the higher yield will result in increased revenues in spite of the overall water supply reductions.

The impact of water policy changes on the nonagricultural economy is another useful policy indicator. Let ψ_i be a regional impact coefficient, denoting an increase in regional product (both direct and indirect effects) associated with a $1.00 increase in revenues of crop *i*. The reduction in regional impact associated with policy scenario $n, \Delta R^n P^n$ is

$$\Delta R^n P^n = \sum_{i=1}^{N}\sum_{j=1}^{J} P_i\left(Y_{ij}^0 - Y_{ij}^n\right)\psi_i. \tag{47}$$

In most cases one expects regional product to decline as a result of reduction in water supply. However, if supply reduction is associated with increased water trading possibilities and higher water prices, regional income may increase because of adoption of conservation technologies that increase yield or increase water used for production of high value crops. These crops generate more revenue per acre-foot of water than low value crops and have stronger linkages to the nonagricultural regional economy due to their higher labor requirements.

The employment impact of a water policy change can also be calculated using standard multipliers. Typically, job loss is measured based on changes in gross revenues. Of course the scope of water trading should mitigate the total labor market impact of water policy changes, particularly if trading results in less high-value fruits and vegetables going out of production following a supply cut.

APPLICATION TO CALIFORNIA

This framework has been applied to study the impact on California agriculture of losing surface water supplies. In the problem considered here, farmers stand to lose between 800,000 and 1,300,000 acre-feet (AF) due to environmental restrictions.

To understand the economic impacts from such a cutback, it is also necessary to know how the cut is to be allocated among users. To a large extent, the final allocation of the supply reduction is an open question, depending on which state or federal agency takes responsibility for the decision. If the state of California makes the decision, then all water users in the state whose consumption affects bay/delta flows are potential targets for cutbacks. However, if the federal government implements the reduced diversions, then only Central Valley Project (CVP) users are liable for the reductions. Thus, the allocation of the cuts is treated as a choice variable, and a variety of initial allocation schemes is considered.

The extent of water trading is currently a policy choice. Trading is highly active within small units such as water districts, and a large volume of water is traded between neighboring districts within the CVP system. There is, however, controversy about how much water can and should be traded among growers, between growers and urban areas, and between basins. Further, there are physical constraints on conveyance that are, at present, hard to define precisely due to hydrological uncertainties and constantly changing regulatory restrictions on pumping. Thus, the scope of the water market is treated here as a policy variable, and the impact models are used to examine a wide array of trading scenarios.

These policy choices are examined using three alternative impact models. These models vary in terms of the production functions, degree of detail, and time scale, as described in Sunding et al. (2001).

BENEFITS TO AGRICULTURE OF WATER TRADING

The incremental costs of removing water from the Central Valley increase sharply as the quantity reallocated increases. Increasing the amount of water devoted to environmental protection from 0.8 million acre-feet (MAF) to 1.3 MAF more than doubles the cost of the regulation to growers. Experimental runs with higher levels of water supply reduction show that this tendency continues and incremental costs of water supply reduction increase as water scarcity increases. This result is attributable to the fact that profit-maximizing farmers will first reduce or cease production of low-value crops in response to reductions in water supply, and will only cease producing high-value crops if the reductions are drastic.

The overall level of the water supply cut is not the most important factor affecting the social cost of protecting bay/delta water quality. Rather, the impacts depend more importantly on the extent of a water market and, when trading is limited, on how supply cuts are distributed among regions. If a market mechanism is used to allocate an annual reduction of 0.8 MAF among a large body of growers in the Central Valley, farm revenue decreases by $10 to $19 million. Using a proportional allocation for the same region, losses are from $45 to $85 million.

COMMENTS

There is increasing pressure in the western United States to protect natural resources by enhancing instream flows. Such policies inevitably mean reducing diversions to irrigated agriculture. This section presents a method for measuring the impacts on agriculture of such reductions. The fundamental tension to be addressed in constructing an agricultural impact model is between the detail necessary to permit examination of the distributional consequences impacts, and the fact that growers have a multidimensional response to policy changes. Rather than constructing a highly complex model incorporating all growing regions and all responses, the results of existing, smaller models can be compared to accurately measure policy impacts in a cost-effective way.

SELECTED BIBLIOGRAPHY

Amemiya, T. 1981. Qualitative response models: A survey. Journal of Economic Literature 19(4):1483-1536.

Anderson, L.T., ed. 1983. Water Rights, Scarce Resource Allocation, Bureaucracy and the Environment. San Francisco, CA: Pacific Institute for Public Policy; Cambridge, MA: Ballinger Publishing Company.

Atkeson, A. and P.J. Kehoe. 1999. Models of energy use: Putty-putty versus putty-clay. American Economic Review 89(4):1028-1043.

Bellon, M.R. and J.E. Taylor. 1993. "Folk" soil taxonomy and the partial adoption of new seed varieties. Economic Development and Cultural Change 41(4):763-786.

Burness, H.S. and J.P. Quirk. 1979. Appropriative water rights and the efficient allocation of resources. The American Economic Review 69(l):25-37.

California Air Resources Board. 1987. Effects of ozone on vegetation and possible alternative ambient air quality standards. Technical Support Document, Sacramento, CA.

Cason, T. N. and R. T. Uhlaner. 1991. Agricultural production's impact on water and energy demand: A choice modeling approach. Resource and Energy Economics 13(4):307-321.

Caswell, M.F. 1983. The diffusion of low-volume irrigation technology in California agriculture. Ph.D. Dissertation, University of California, Berkeley, CA.

Caswell, M.F. and D. Zilberman. 1985. The choices of irrigation technologies in California. American Journal of Agricultural Economics 67(2):223-234.

————. 1986. The effects of well depth and land quality on the choice of irrigation technology. American Journal of Agricultural Economics 68(4):798-811.

Chakravorty, U., E. Hochman, and D. Zilberman. 1995. A spatial model of optimal water conveyance. Journal of Environmental Economics and Management 29(l):25-41.

Coase, R. 1992. The institutional structure of production. The American Economic Review 82(4):713-719.

County Supervisors Association of California (CSAC). 1990. California County Fact Book 1988-1989. Sacramento, CA.

Davies, S. 1979. The Diffusion of Process Innovations. Cambridge, UK: Cambridge University Press.

Dinar, A. and D. Yaron. 1992. Adoption and abandonment of irrigation technologies. Agricultural Economics 6(April):315-332.

Dinar, A. and D. Zilberman. 1991. Effects of input quality and environmental conditions on selection of irrigation technologies. In: A. Dinar and D. Zilberman, eds., The Economics and Management of Water and Drainage in Agriculture. Norwell, MA: Kluwer Academic Publishers.

————-1991b. The economics of resource-conservation, pollution-reduction technology selection: The case of irrigation water. Resource and Energy Economics 13(4):328-348.

Dinar, A., S. Hatchett, and E. Loehman. 1991. Modeling regional irrigation decisions and drainage pollution control. Natural Resource Modeling 5(2):191-212.

Dixon, L., N. Moore, and S. Schechter. 1993. California's 1991 drought water bank. RAND Corporation, Santa Monica, CA.

Domencich, T. and D. McFadden. 1975. Urban Travel Demand: Behavioral Analysis. Amsterdam: North-Holland Publishing.

Fudenberg, D. and J. Tirol. 1991. Game Theory. Cambridge, MA: MIT Press.

Gisser, M. and R.N. Johnson. 1983. Institutional restrictions on the transfer of water rights and the survival of an agency. In: T.L. Anderson, ed., Water Rights, Scarce Resource Allocation, Bureaucracy, and the Environment. San Francisco, CA: Pacific Institute for Public Policy; Cambridge, MA: Ballinger Publishing Co.

Gisser, M. and D.A. Sanchez. 1980. Competition versus optimal control in ground water pumping. Water Resources Research 16(4):638-642.

Goodman, C. and R.E. Howitt. 1986. The California agricultural resources model: Structure, calibration and applications under positive quadratic programming. Working Paper, Department of Agricultural Economics, University of California at Davis.

Green, G. and D. Sunding. 1997. Land allocation, soil quality and irrigation technology choices. Journal of Agricultural and Resource Economics 27(2):367-375.

————. 2000. Designing environmental regulations with empirical microparameter distributions: The case of seawater intrusion. Resource and Energy Economics 22(1):63-78.

Green, G., D. Sunding, D. Zilberman, and D. Parker. 1996. Explaining irrigation technology choice: A microparameter approach. American Journal of Agricultural Economics 78(4): 1064-1072.

Greene, D.L., J.R. Kahn, and R.C. Gibson. 1999. Fuel economy rebound effect for U.S. household vehicles. Energy Journal 20(3):1-31.

Hall, D.C. and M.W. Hanemann. 1996. Urban water rate design based on marginal cost. In: D.C. Hall, ed., Advances in the Economics of Environmental Resources, Volume 1. Greenwich, CT: JAI Press, Inc., pp. 95-122.

Hausman, J.A. 1979. Individual discount rates and the purchase and utilization of energy using durables. Bell Journal of Economics 10(l): 33-54.

Hicks, J.R. 1939. Value and Capital. Oxford: Clarendon Press.

Hochman, E. and D. Zilberman. 1978. Examination of environmental policies using production and pollution microparameter distributions. Econometrica 46(4):739-760.

Houthakker, H. 1955-56. The Pareto distribution and the Cobb-Douglas production function in activity analysis. Review of Economic Studies 23(l):27-31.

Howe, W.C., D.R. Schurmeier, and W. Douglas. 1986. Innovative approaches to water allocation: The potential for water markets. Water Resources Research 22(4):439-445.

Howitt, R.E. 1989. The economic assessment of California field crop losses due to air pollution. Working Paper, Department of Agricultural Economics, University of California at Davis.

————.1992. Economic assessment of acid deposition and ozone damage on the San Joaquin valley agriculture. Working Paper, Department of Agricultural Economics, University of California at Davis.

————. 1995. Positive mathematical programming. American Journal of Agricultural Economics 77(2):329-342.

Johansen, L. 1959. Substitution versus fixed production coefficients in the theory of economic growth: A synthesis. Econometrica 27(April):157-176.

————. 1972. Production Functions: An Integration of Micro and Macro, Short-Run and Long-Run Aspects. Amsterdam: North Holland.

Just, R., D. Hueth, and A. Schmitz. 1982. Applied Welfare Economics and Public Policy. Englewood Cliffs, NJ: Prentice Hall.

Knapp, K.C., A. Dinar, and J. Letey. 1986. On-farm management of agricultural drainage problems: An economic analysis. Hilgardia 54(4):1-31.

Letey, J. and A. Dinar. 1986. Simulated crop-water production functions for several crops when irrigated with saline waters. Hilgardia 54(l):1-32.

Letey, J., A. Dinar, and K. Knapp. 1985. A crop-water production function model for saline irrigation waters. Soil Science of America Journal 49(4):1005-1009.

Lichtenberg, E. 1989. Land quality, irrigation development, and cropping patterns in the northern high plains. American Journal of Agricultural Economics 71(l):187-194.

Moreno, G. and D. Sunding. 2001. Factor price risk and conservation technology adoption. Working Paper, Department of Agricultural and Resource Economics, University of California at Berkeley.

Moreno, G., D. Osgood, D. Sunding, and D. Zilberman. 1999. The economics of water supply reliability. Working Paper, Department of Agricultural and Resource Economics, University of California at Berkeley.

Moyle, P. and R. Yoshiyama. 1992. Fishes, aquatic diversity management areas, and endangered species: A plan to protect California's native aquatic biota. California Policy Seminar, University of California at Berkeley.

Negri, D. and D. Brooks. 1990. Determinants of irrigation technology choice. Western Journal of Agricultural Economics 15(2):213-223.

Olmstead, J., D. Sunding, D. Parker, R. Howitt, and D. Zilberman. 1997. Water marketing in the '90s: Entering the electronic age. Choices (Third Quarter):15-19.

Perrin, R.K. and D. Winkelmann. 1976. Impediments to technical progress on small versus large farms. American Journal of Agricultural Economics 58(5):888-894.

Rausser, G. and P. Zusman. 1991. Organizational failure and the political economy of water resources management. In: A. Dinar and D. Zilberman, eds, The Economics and Management of Water and Drainage in Agriculture. Norwell, MA: Kluwer Academic Publishers.

Rosen, D.M. 1989. Property rights and public choice in water districts: An application to water markets. Ph.D. Dissertation, University of California at Berkeley.

Rothschild, M. and J. Stiglitz. 1970. Increasing risk 1: A definition. Journal of Economic Theory 2(3):225-243.

Scheuring, A.F., ed. 1983. A Guidebook to California Agriculture by Faculty and Staff of the University of California. Berkeley: University of California Press.

Shrestha, R. and C. Gopalakrishnan. 1993. Adoption and diffusion of drip irrigation technology: An econometric analysis. Economic Development and Cultural Change 41(2):407-418.

Smith, T. Rodney. 1984. The economic determinants and consequences of private and public ownership of local irrigation facilities. In: T.L. Anderson, ed., Water Rights, Scarce Resource Allocation, Bureaucracy, and the Environment. San Francisco, CA: Pacific Institute for Public Policy; Cambridge, MA: Ballinger Publishing Co.

State of California, Department of Water Resources. 1980. Measuring Economic Impacts, the Application of Input-Output Analysis to California Water Resources Problems. State of California, The Resources Agency, Sacramento, CA.

————. 1985. CIMIS Final Report. State of California, The Resources Agency, Sacramento, CA.

————. 1989. Drought Contingency Planning Guidelines for 1989. State of California, The Resources Agency, Sacramento, CA.

————. 1991. California's Continuing Drought 1987-1991. State of California, The Resources Agency, Sacramento, CA.

————. 1993. California Water Plan Update. State of California, The Resources Agency, Sacramento, CA.

Stoneman, P.L. and N. Ireland. 1986. Technological diffusion, expectations and welfare. Oxford Economic Papers 38(2):283-304.

Sunding, D., D. Zilberman, and N. MacDougall. 1995. Water markets and the cost of improving environmental quality in the San Francisco Bay/delta estuary. West-Northwest Journal of Environmental Law and Policy 2(3):159-165.

Sunding, D., D. Zilberman, R. Howitt, A. Dinar, and N. MacDougall. 2001. Measuring the costs of reallocating water from agriculture: A multi-model approach. Natural Resources Modeling: in press.

Tsur, Y. and A. Dinar. 1995. Efficiency and equity considerations in pricing and allocating irrigation water. World Bank, Policy Research Working Paper No. 1460, Washington, DC.

Wichelns, D. 1991a. Increasing block-rate prices for irrigation water motivates drain water reduction. In: A. Dinar and D. Zilberman, eds., The Economics and Management of Water and Drainage in Agriculture. Norwell, MA: Kluwer Academic Publishers.

———. 1991b. Motivating reductions in drain water with block-rate prices for irrigation water. Water Resources Bulletin August.

Wirl, F. 2000. Lesson from utility conservation programs. Energy Journal 21(1):87-108.

Zilberman, D. and F. Shah. 1994. Queuing vs. markets in water resource allocation. Working Paper, Department of Agricultural and Resource Economics, University of California at Berkeley.

Zilberman, D., A. Dinar, N. MacDougall, M. Khanna, C. Brown, and F. Castillo. 1995. Individual and institutional responses to drought: The case of California agriculture. Working Paper, Department of Agricultural and Resource Economics, University of California at Berkeley.

Zusman, P. 1988. Individual Behavior and Social Choice in A Cooperative Settlement. Jerusalem: Hebrew University, Magnes Press.

———. 1991. A conceptual framework for a regulatory policy of the Israeli water resources. Working Paper No. 9108, The Center For Agricultural Economic Research, Hebrew University, Jerusalem.

Water Allocation and Pricing in Agriculture of Iran

Abbas Keshavarz, Shahram Ashraft[1], *Nader Hydari*[1], *Morteza Pouran*[1], *and Ezzat-Allah Farzaneh*[1]

ABSTRACT

Limited water resources and world population growth have caused a shortage of agricultural products in some countries. Currently, this limitation is one of the most serious problems in Middle Eastern countries, especially arid and semi-arid countries. Considering population growth rates and limited water resources in the world, it is anticipated that food security will be a serious challenge in coming decades.

The agricultural sector in Iran is one of the most important economic sectors of the country, and water is the most limiting factor for production. More than 90 percent of the renewable water in the country is used in agriculture, but its production is insufficient to meet the country's demand. Because of low irrigation efficiency, about 50 to 60 percent of renewable water is lost in agriculture, and this has caused agricultural water productivity to be very low.

Efficient application of water in agriculture is one of the most important contributing factors to producing as much food as is required at present and in the future. Proper planning, management, and education in this sector would help prevent the waste of limited natural resources. Efficient application of water in the country, therefore, is one of the most important policies of the government of the Islamic Republic of Iran (I.R. Iran).

---

[1] Scientists of Iranian Agricultural Engineering Research Institute.

INTRODUCTION

Iran is located in the Northern Hemisphere, between 25° and 40° N and 44° to 63° E. Agriculture plays an important role in the economy of Iran. It accounts for one fourth of the Gross Domestic Product (GDP), one fourth of employment, more than 80 percent of food requirements, one third of non-oil exports, and 90 percent of raw materials for industries. The agriculture of I.R. Iran enjoyed an average growth rate of 5.1 percent over the two National Development Plans (1989 to 1999).

Out of the 165 million hectares that comprise the country's area, about 37 million hectares are suitable for irrigated and dryland agriculture, of which 20 million hectares are irrigated and 17 million hectares are dryland. Of the 37 million hectares of agricultural lands, currently 18.5 million hectares are devoted to horticulture and field crop production. Of these, 6.4 million hectares are under annual irrigated crops, 2 million hectares are under horticultural crops, and about 6.2 million hectares are under annual dryland crops, while the remaining 3.9 million hectares are fallow.

The total natural resources and rangeland areas are about 102.4 million hectares, composed of 90 million hectares as pastures (in various level of forage productivity) and 12.4 million hectares as forests.

CLIMATE, RAINFALL, AND EVAPORATION

The Islamic Republic of Iran is situated in one of the most arid regions of the world. The average annual precipitation is 252 mm (one-third of the world's average precipitation), and this is under conditions in which 179 mm or 71 percent of rainfall is directly evaporated. The annual evaporation potential of the country is between 1500 and 2000 mm. Unfortunately, in the past six years, particularly in the year 2000, some parts of the country have suffered severely from drought.

Altitudes vary from 40 m below sea level to 5,670 m above sea level and have a pronounced influence on the diversity and variation of the climate. Although most parts of the country can be classified as arid to semiarid, the country enjoys a wide range of climatic conditions. Both latitude and altitude have a major influence on climate in the various regions. This can be seen in the geographic variation of annual precipitation (from 50 mm in the central desert to 1600 mm in Gillan Province, situated at the southern coast of the Caspian Sea), and a wide range of temperatures that can vary up to 100°C (from –44°C in Borudjen/ Chahar Mahal Bakhtiari Province, located in the central Zagrus Range mountains to 56°C in the south along the Persian Gulf coast). Distribution of precipitation in Iran is presented in Table 1.

TABLE 1 Distribution of Precipitation in Iran

Annual precipitation (mm)	Area (km^2)	Percent
<50	100,000	6
50-100	285,000	17
100-200	456,000	28
200-300	370,000	23
300-500	280,000	17
500-1000	130,000	8
>1000	18,000	1
Total	1,648,000	100

WATER RESOURCES AND WATER USE

The main source of water in Iran is precipitation of both rainfall and snow (70 percent rainfall and 30 percent snow). Total precipitation is estimated to be about 413 billion cubic meters (bcm), of which about 71.6 percent (295 bcm) directly evaporates. By taking into account 13 bcm of water entering from the borders (joint border rivers), the total potential renewable water resources have been estimated to be 130 bcm.

Currently, the total water consumption is approximately 88.5 bcm, out of which more than 93 percent is used in agriculture, while less than 7 percent is allocated for urban and industrial consumption (Table 2).

Under the present situation, 82.5 bcm of water are utilized for the irrigation of 8.4 million hectares of irrigated agriculture (horticulture and field crops). About 1.4 million hectares of these areas are managed by regulated flow (irrigation networks); 6.7 million hectares by means of traditional networks, and less than 300,000 hectares under a pressurized system.

Surface water resources provide 37.5 bcm of water for different consumption purposes (about 42 percent of the total water consumed) in the country. The existence and importance of groundwater has been known and understood for

TABLE 2 Estimated Consumption of Water in Iran (Year 1998)

Consuming Sector	Consumption (bcm)	Percent of total
Agriculture	82.5	93.22
Urban	5.6	6.32
Industry	0.03	0.03
Miscellaneous	0.37	0.43
Total	88.5	100

thousands of years. The traditional method of groundwater extraction is Qanat, which brings water to the surface by gravity. In recent years more than 50,000 wells of various types have been dug and used for extraction of groundwater from the aquifers. More than 60 percent of total water consumption in the country (51 bcm) is extracted from groundwater resources. Due to inefficiency of traditional irrigation methods and water conveying systems, about 63 percent of the valuable water is lost, and in practice only 37 percent of available water is utilized in agricultural production.

SOCIOECONOMIC ASPECTS

Agriculture in Iran is run privately, with 99 percent of the agricultural lands being managed and commodities being produced by the private sector.

Out of a total population in Iran of 65 million, about 24 million (38 percent) live in rural areas (roughly 60,000 villages). Thirty years ago, in the 1970s, the rural to urban ratio was almost the reverse. Between 1976 and 1998, the rural population increased from 17.9 million to 23.6 million. Out of 14.2 million job opportunities throughout the country, 3.3 million (23 percent) are in the agricultural sector. The distribution of the rural population is largely determined by the availability of water, rainfall, and arable lands. Thus, with the exception of the well-watered Caspian Sea area with its high rainfall, most settlements are isolated and scattered through arid and semiarid regions in the plains and mountains. More than 90 percent of the 2.6 million rural agricultural households possess land, and a great majority of these are small and medium farms, which dominate Iranian agriculture.

A high proportion of farms in Iran are considered small in size. About 78 percent of farmers have farms of less than 10 hectares in size, and 11 percent are less than one hectare. While farms less than 10 hectares make up about 37 percent of the cultivated land, represent about 12 million of the rural population, and produce a similar proportion of the agricultural gross output, they produce less than 10 percent of marketed agricultural production. This is accounted for by the fact that small farmers have low income, and many of them are mainly subsistence farmers with no surplus products for sale. Farms over 10 hectares provide about three-quarters of the market supplies. Large farms (over 100 hectares), currently only make small contributions to the agricultural output of Iran.

The national farm-size distribution for irrigated and dryland agriculture in Iran is presented in Table 3. Farms less than 10 hectares account for a greater percentage of the irrigated areas.

AGRICULTURAL PRODUCTION IN IRAN

The production of horticulture and field crops in the decade from 1988 to 1998 experienced positive changes. These changes were due mainly to the imple-

TABLE 3 National Farm-size Distribution (Irrigated and Dryland Agriculture)

Size (ha)	Total Area (kha)	Area (percent)	Percentage Irrigated	Dryland	Average Size (ha)
< 1	196	1.6	3.1	0.3	0.4
1-2	423	3.4	5.7	1.4	1.1
2-5	1,630	13.0	17.4	9.3	2.4
5-10	2,371	18.9	20.4	17.6	4.8
10-25	4,467	35.5	29.1	40.7	9.8
25-50	1,585	12.6	10.0	14.7	21.2
5-100	961	7.5	6.2	8.8	41.8
Over 100	947	7.5	8.0	7.1	118.5
Total	12,580	100	100	100	4.9

mentation of the results of research findings and extension activities that contributed to increased yield per unit area. The other factor that caused change was utilization of unused land capacities through development and expansion of areas under cultivation. The average annual growth rate of agricultural production was 5.1 percent during the period from 1988-1998. The total agricultural production for various commodities from 1988-1998 is presented in Table 4.

During the same period wheat production increased from 7.2 million tons to 11 million tons, corn grain from 0.143 million tons to 0.941 million tons, and vegetables and horticulture crops from 5.4 million tons to 11.6 million tons. It is noteworthy that the 1998 growing season was an exceptional year in terms of both the amount and distribution of precipitation. Unfortunately, in the last three

TABLE 4 Agricultural Production in Various Commodities during Years 1988-1998

Agricultural Products	1988 (in million tons)	1998 (in million tons)
Field Crops	28.60	53.30
Irrigated	25.85	44.73
Dryland	2.75	8.57
Horticultural Crops (Fruits)	7.20	11.60
Milk	3.40	5.10
Red Meat (lamb and beef)	0.51	0.75
Chicken	0.30	0.70
Eggs	0.26	0.50
Fish	0.24	0.40

years (1999-2002), particularly in 2002, drought has seriously threatened some parts of the country, and agricultural production decreased about 8-10 percent. Annual crop groups and horticultural crops and their areas under cultivation and production are presented in Tables 5 and 6, respectively.

TABLE 5 Cultivated Areas and Total Production of Different Crops under Irrigated and Dryland Conditions in 1998 in Iran

	Cultivated areas (kilohectares)			Total Production (kilotons)		
Crop Groups	Irrigated	Dryland	Total	Irrigated	Dryland	Total
Cereals[a]	3,706	5,069	8,775	13,453	5,513	18,967
Pulses[b]	173	785	957	250	327	577
Industrial Crops[c]	590	132	751	7,629	140	7,769
Vegetables[d]	423	35	458	9,730	428	10,158
Summer Crops[e]	288	47	335	5,316	277	5,593
Forage Crops[f]	786	152	938	8,899	1,179	10,078
Total annual crops	5,966	6,220	12,186	—	—	—
Horticulture (orchards)	1,777	188	1,965	11,397	263	11,660
Total	7,797	6,501	14,296	56,796	8,176	64,073

[a]Cereals: wheat, barley, paddy, and corn grain.
[b]Pulses: pea, bean, lentil, Fava bean, and other pulses.
[c]Industrial crops: cotton, sugar beet, sugarcane, and oil seeds.
[d]Vegetable crops: potato, onion, tomato, eggplant, etc.
[e]Summer crops: melon, watermelon, cucumber, etc.
[f]Forage crops: alfalfa, clover berseem, and forage corn.

TABLE 6 Cultivated Areas and Total Production of some Horticultural Crops in 1998

Crops	Cultivated Areas (kilohectares)	Production (kilotons)
Apple	163	2,137
Grape	288	2,342
Pistachio	361	131
Citrus	234.5	3,934
Date Palm	215.7	908
Pomegranate	55	604
Tea	34	270
Nuts[a]	217	250
Olive	30	24
Others[b]	403	987
Total	2,000	11,600

[a]Almond, Walnut, Hazelnut.
[b]Pear, Peach, Cherry, Apricot, Strawberry, Kiwifruit, Fig, Saffron, etc.

EVALUATION OF WATER USED IN AGRICULTURE

Water use in agriculture in Iran may be evaluated in terms of irrigation efficiency and agricultural water productivity.

Irrigation Efficiency

First of all the authors acknowledge that the discussions in this section are based on personal communication with irrigation experts and research studies in the country that might differ in some ways from official reports.

A common perception is that irrigation water is wasted considerably in Iran. Studies have shown that the overall irrigation efficiency in Iran ranges from 33 percent to 37 percent, which is lower than average worldwide irrigation efficiency. This rate of irrigation efficiency indicates that the average consumption of irrigation water in the country is high compared to worldwide consumption of irrigation water. Comparisons between the application of irrigation water in Iran and worldwide application for different crops are shown in Table 7.

In 1999 the Iranian Agricultural Engineering Research Institute (IAERI) conducted a research project on the on-farm water application efficiency for different crops. The results obtained from field experiments in different provinces showed that the application efficiency depends on farm management, method of irrigation, growth stage, and type of crop, and it varied from 24.7 percent to 55.7 percent. Considering the conveyance efficiency in the target area of the research study, the overall irrigation efficiency varied between 15 and 36 percent.

Based on results obtained from studies and research conducted by different water organizations for determining overall irrigation efficiency, IAERI has published the concerned national document in 2000 (Dehghani S.H, A. Alizadeh and A. Keshavarz, 1999). The overall irrigation efficiency in different provinces in the country is presented in Table 8. As is shown in this table, there is a range of overall irrigation efficiencies for each province; the irrigation efficiency varies from a low to a high value based on the source of water and type of irrigation

TABLE 7 Average Application of Water for Different Crops Worldwide and in Iran (1994)

Crops	World (m^3/ha)	Iran (m^3/ha)
Wheat	4,500-6,500	6,400
Melons	7,000-10,500	17,900
Sugar beet	5,500-7,500	10,000-18,000
Rice	4,500-7,000	10,000-18,000
Sugarcane	15,000-25,000	20,000-30,000
Corn	5,000-8,000	10,000-13,000

TABLE 8 Ranges of Irrigation Efficiency in some Provinces in Iran

No	Provinces	Range of Irrigation Efficiency (%)
1	West Azerbaijan	28-41
2	Ardabil	28-39
3	Isfahan	28-42
4	Boshehr	24-30
5	Chaor-Mahhal-Bottiari	30-39
6	Korasan	30-37
7	Kozestan	27-37
8	Zanjan	25-38
9	Semnan	30-40
10	Gazvin	27-38
11	Kordestan	25-40
12	Glolestan	28-40
13	Gillan	38-54
14	Mazanderan	37-57
15	Markazi	29-39
16	Hamedan	27-38
17	Yazd	30-40

management applied for the distribution of water to the farms. The source of water may be surface water, shallow wells, qanats, and springs, and the distribution networks may be modern or traditional irrigation networks. Considering overall irrigation efficiencies in different provinces (Table 8) and different published reports and research studies, the irrigation efficiency in the country is estimated to be 35 percent, which is very low compared to developing countries (45 percent) and developed countries (60 percent).

According to published agricultural statistics, the areas under irrigation in Iran total 8.4 million hectares (Mha). For irrigation of these areas, 82.5 bcm of water is used. In contrast, the international water management institute (IWMI) has reported that the average net irrigation requirement in Iran for cereal and field crops is 5,100 and 8,100 cubic meters per hectare (m^3/ha), respectively. The Ministry of Energy, which is in charge of water allocation in Iran, has estimated the average amount of irrigation requirement to be 5,200 cubic meters per hectare. The average of figures that have been published by different consulting engineers is 5,900 m^3/ha. Based on these figures, the overall irrigation efficiency in Iran is between 48 and 55 percent, which is quite different from the figures that are presented officially or unofficially by various sources. Based on available data, the areas that receive full irrigation in Iran total 5 Mha. At least 1.6 and 1.8 Mha of the irrigated areas are suffering from severe and moderate water stress, respectively.

Half of the fully irrigated areas are equipped with modern irrigation systems and operated by governmental organizations. Irrigation efficiency in such systems is very low and is measured at 20-30 percent. Irrigation efficiency may be so low due to the free availability of water released from dams, there being no incentive for farmers to save water. The other half of fully irrigated areas are operated by the private sector, and the water is supplied from groundwater resources. In this case, the irrigation efficiency is also rather low and has been measured to be about 35 percent. The rest of the irrigated farms in Iran, which are under severe to moderate water stress, belong to small farm holders who do not save water, but their irrigation efficiency is quite high. Irrigation efficiency in these farms is estimated to be 55-65 percent. The reason for this may be due mainly to reduced irrigation, which they usually practice. These farmers have enough land, and the area under their cultivation is much greater than the water available to them. They usually receive more benefit from their extensive farming with reduced irrigation in comparison with those who practice intensive farming and full irrigation.

AGRICULTURAL WATER PRODUCTIVITY

Water productivity is simply defined as the amount of production per unit of water. Water productivity can be increased by obtaining greater production with the same amount of water, or by reallocating water from lower to higher value crops or from one section of the agriculture sector to another, where the marginal value of the water is higher.

Indeed, the greatest increases in productivity of water in irrigation have not been from better irrigation systems, but rather from increased crop yields due mainly to better management. A key to mitigating the problem of water scarcity in Iran is increasing the productivity of water. Let us consider agriculture, as there is a tremendous need to produce food, mainly grains, from water resources. There are, generally, four approaches for generating more agricultural output from utilizable water resources as follows:

1. increasing utilizable water,
2. developing more primary water (increase in development of facilities),
3. consuming more of the developed water efficiently (increase in basin efficiency), and
4. producing more output per unit of water consumed (increase in water productivity).

Opportunities to increase agricultural water productivity via the first two approaches are very limited in Iran. For the third approach, it may be feasible to improve water conservation in those areas that receive full irrigation. By proper water management and completion of water distribution networks, irrigation efficiency is expected to increase by 50 to 60 percent. As a result, the areas of these lands may increase from 2.5 Mha to 3.5 and 4.3 Mha, respectively. Although it would also be possible to increase irrigation effectiveness in those areas that receive reduced irrigation, no significant increase in cultivated areas could be expected. Thus, the major solution will be the fourth approach, which is to increase water productivity. This is especially appropriate in the areas where water application is much higher than water requirements of the crops, as estimated by the soil water balance during the growing season in different field studies.

Crop water productivity (CWP) can be defined as the kilograms (kg) of yield per unit of consumed water (ET), or CWP = kg/ET. An extensive data set of studies carried out worldwide (Doorenbos and Kassam, 1979) has shown that CWP is very low for some crops, like sunflower, but for other crops like tomato it is generally high. The genetic characteristics of the crop are the primary factor determining CWP. A secondary factor that affects CWP in various ways is the reaction to water deficit. For instance, CWP for wheat in Khorasan Province of Iran is about 0.5 kg/m^3, which is quite low compared to a similar environment like the Imperial Valley in California or even Bhakra in India.

The CWP for different agricultural crops in Iran is shown in Table 9. As a rule of thumb, a reasonable level of wheat productivity is about 1 kg/m^3. Therefore, if the demand for grain grows by 50 percent in the country by 2020, one way to meet this demand is to increase water productivity by 50 percent.

DEMAND FOR FOOD AND AGRICULTURAL PRODUCTS

Studies have shown that despite population increases from 1988-1998, the agricultural production per capita index has increased remarkably to 137 per-

TABLE 9 Measured Crop Water Productivity at Different Regions in Iran

Crop	Irrigation Method	Yield (kg)	Applied Water (m^3/ha)	Water Productivity (kg/m^3)
Wheat	Furrow-border	5,460	9,900	0.55
Barley	Furrow	6,090	6,120	1.00
Sugar beet	Furrow-border	37,700	14,500	2.60
Potato	Furrow	37,100	5,140	7.21
Corn	Furrow	7,000	1,080	0.65
Alfalfa	Basin-border	10,500	11,660	0.90
Beans	Furrow	5,100	5,600	0.91
Sesame	Furrow	1,432	7,000	0.20
Tomato	Furrow	16,000	4,800	3.33
Lettuce	Furrow	4,100	8,600	4.77

cent. The agriculture sector has produced 80 percent of the food supply in the past decade. The energy supply per capita from 1988-1998 was 2,900 calories. According to international standards, the daily energy requirement is 2,300 calories per capita. Food supply in Iran consists of 72 percent carbohydrates, 17.5 percent fat, and 10.4 percent protein. In comparison, a desirable composition is 55-75 percent carbohydrates, 15-30 percent fat, and 15-20 percent protein.

About 85 percent of energy and 76 percent of protein in the Iranian diet is provided by 10 food commodities, which include bread, rice, vegetable oil, sugar, chicken, pulses, milk and its by-products (cheese and yogurt). These are called basic commodities in the Iranian diet. Around 65 percent of total food expenditures are spent on these items. Bread alone provides 40 percent of the energy and 45 percent of the protein in the Iranian diet; therefore, wheat is the staple component of Iranian daily food. However, improvements in the food consumption pattern should be made in coordination with the domestic production potential.

SHORT TERM AGRICULTURAL SCENARIO

The agriculture sector supplies most of the food requirements in Iran; however, it is not yet self sufficient. The average quantity of major commodities that were imported annually, during 1988–1998, are presented in Table 10.

Iran's population is expected to increase from about 63 million in 2000 to 70 million people by 2005. Challenges in the future include domestic food supply as well as improvement in food consumption patterns. In the short term, which coincides with the Third Five-Year Development Plan (2001-2005), objectives and highlights of the agriculture program can be summarized as follows:

- Improving quality and quantity of agricultural products (annual crops, fruits, livestock, poultry, fisheries, and aquaculture) through efficient and sustainable use of natural resources, while approaching food security. Priorities are given to strategic crops, including cereals, oil seeds, olives, maize, forage crops, fisheries, and export extension for horticultural products.

TABLE 10 Major Commodities Imported During 1988-1998 (Annual Average)

Commodity	Amount	Unit
Wheat	3.4	million tons
Rice	600	kilotons
Barley	500	kilotons
Maize	1.2	million tons
Sugar (raw)	840	kilotons
Oil crops	750	kilotons
Meat	520	kilotons

- Drafting and completion of all food safety related standards to improve the health status of society.
- Implementing various policies, including determination of a suitable model for sustainable development. This is planned to be developed based on current needs and capacities and on comparative advantages of increasing yields through efficient use of land and water resources. Innovative approaches to increase use of certified seeds, optimize application of pesticides and fertilizers, and expand development of mechanization are other aspects of these policies.
- Improving activities related to soil moisture conservation and retention, watershed management, and protection of agricultural land. The geographic location of Iran makes it very vulnerable in terms of its annual precipitation. This factor makes utilization of dryland agriculture with low yields very difficult. The highest priority, therefore, is introduction of drought resistant varieties of crops for irrigated and dryland areas to increase water use efficiency and water productivity.

Agricultural products are expected to increase from 60 million tons in 2001 to 90 million tons by 2004. Currently, total production of irrigated crops is 57 million tons, while total water supply for agriculture is 83 bcm. Therefore, water productivity is 0.7 kg/m^3. By the year 2005, total production of irrigated crops is expected to reach 90 million tons, while total water supply will be about 90 bcm so projected water productivity is estimated to be 1 kg/m^3 of water.

CHALLENGES IN THE 21ST CENTURY

By 2020 Iran's population is estimated to reach 100 million. However, total agricultural production is expected to be 200 million tons, of which 189 million tons will be harvested from irrigated crops. Total water supply for agriculture will be about 100 bcm. This means that by the year 2020, water productivity should reach 1.9 kg/m^3. To fulfill this expectation, agricultural commodities (wheat, barley, maize, oil crops) will be the main focus for improvements. With opportunities for expanding areas under cultivation almost exhausted, additional food production will have to be accomplished mainly through increasing productivity. Demand for major agricultural products by the year 2020 is presented in Table 11.

Intensive use of water, fertilizers, and other agricultural inputs for crop production at present are the major cause of problems in soil and groundwater salinization, nutrient imbalances, incidence of new pests and diseases, and environmental degradation.

Rising biotic pressure, lack of a suitable soil management system, and lack of inputs to realize optimum potential of land appear to threaten sustainability of agriculture. Thus, the consequences are degraded lands, loss of biodiversity, soil

TABLE 11 Demand for Major Agricultural Products, Comparing Production in 1998 with Projected Demand in 2020

Commodity	Production (1988) (million tons)	Demand (2020) (million tons)
Wheat	10.00	17.50
Rice (Paddy)	2.70	4.90
Maize	0.90	4.30
Pulses	0.68	1.75
Oil seed	0.24	3.80
Potato	3.30	14.00
Forage crop	7.98	24.10
Apple	2.10	12.00
Grape	2.34	5.70
Pistachio	0.13	0.75
Citrus	3.90	7.30
Date-palm	0.91	3.00
Tea	0.27	0.40
Olive	0.02	1.30

erosion, deforestation, and overall environmental pollution, all of which result in lowered productivity. For efficient and sustainable agriculture it will be essential to shift from a commodity-centered approach to a farming-systems approach, which calls for multidisciplinary efforts. This will require emphases on efficiency, sustainability, post-harvest management, mechanization, marketing, and trade. Such an approach will also require forging links with all who are concerned at the regional, national, and international levels.

With implementation of these new approaches, water productivity should be increased to at least 1.9 kg/m^3. This implies that the institutional structure and procedures of water allocation in the agriculture sector should be modified. This would necessitate emphases on special prioritization, policies, modernization, water use efficiency, and productivity management.

To elaborate more on the importance and role of water, and to draw the necessary attention for improving water productivity in the future, the main agricultural criteria in an index year (2000) and short and long horizons are presented in Table 12.

As shown in Table 12, the possible increase in water resources is very limited (9.7 and 22 percent after 5 and 22 years, respectively). However, in order for agricultural products from the irrigated land to increase significantly (by 150 and 337 percent by 2005 and 2020, respectively), water productivity (0.7 kg/m^3) has to be increased to 1 and 1.9 kg/m^3 by 2005 and 2020, respectively.

TABLE 12 Main Agricultural Criteria of Iran at Index year (2000) and Short and Long Horizons

Criterion	Year 2000	2005	2020
Population (million)	63	70	100
Volume of water allocated to the agriculture sector (billion cubic meter)	82.5	90	100
Total types of products of irrigated lands	56	85	189
Water productivity (kg/m^3)	0.7	1	1.9
Increase in allocated water (%)	—	9.7	22
Expected increase in the products of irrigated lands (%)	—	150	337
Percent of total water allocated to the agriculture sector	93	93	93

OBJECTIVES AND CHALLENGES

As shown in Table 12, it is anticipated that agricultural products from irrigated areas will increase from 56 million tons in 2000 to 85 million tons in 2005 and 189 million tons in 2020. This would be realized when our water resources can be increased up to a maximum of 10-22 percent. Therefore, it is necessary to increase water productivity in agriculture from 0.7 kg/m^3 in 2000 to 1.9 kg/m^3 by 2020. The expected increase in agricultural products basically depends on the country's available water resources.

Water scarcity is the most limiting factor in agricultural productivity in Iran. Attention to improvements in water supply and water productivity programs has been the most important and governing policy during the past 22 years. Under this policy, different rules have been applied and, in addition, different technical infrastructures (executive, research, and consultative) in both public and private sectors have been developed. This attention, in addition to establishment of special laws and regulations, has been considered in the construction of development programs.

Among the established laws which can be nominated are the Balanced Distribution of Water law (established in 1983) and the Executive Instructions of Optimization of the Agricultural Water Consumption. The above mentioned law is a focal point in the evolution of viewpoints on water issues in I.R. Iran. The basic objectives of this law are as follows:

- Optimum use of water resources in agriculture through optimizing consumption.
- Volumetric distribution of water based on irrigation requirements in the different regions.

- Introduction of suitable cropping patterns (relative to water productivity).
- Improvement of irrigation management.

The first step in fulfillment of these objectives is determination of water requirements of different crops in the different regions. Provision of suitable cropping patterns, and determination of expected irrigation efficiency in different time periods would be the next step. Carrying out these steps in terms of volumetric water distribution and consumption, and developing suitable water pricing policies in local and regional operation and management of irrigation networks, have as their aim the optimization of agricultural water consumption. The following sections summarize actions taken to optimize agricultural water consumption. Many of these actions are documented as the National Water Document, certified by the government of Iran (Ministry of Agriculture, 1998).

ESTIMATION OF CROP WATER REQUIREMENTS

Determination of the water requirement of crops is the basic measure in irrigation and water resource planning. In the past 50 years widespread and extensive research has been conducted worldwide, and as a result more operational as well as accurate procedures for estimation of crop water requirements are available. These research findings have contributed to better recognition of the physical and biological processes of evapotranspiration. On the basis of these findings, models and practical procedures for estimating crop water requirements have been developed. From the wide variety of studies carried out using different methods and under various climatic conditions throughout the world, it is necessary to select those that have the greatest applicability to the climate and geophysical conditions in Iran.

There has been much discussion about which method of estimating evapotranspiration of field crops (ETC) and crop water requirements is the most accurate. The results of a study to determine irrigation and water requirements conducted by the American Society of Civil Engineers (ASCE) and the Society of European Commission in a consortium of research institutes have shown the superiority of the Penman-Monteith method over other methods for the estimation of crop water requirement. These results have shown that the Penman-Monteith method (Monteith, 1965, and Penman, 1948) with a 4 percent overestimate in wet regions and 1 percent underestimate in dry regions (compared with lysimetric data), is the best method for ETC estimation.

The Penman-Monteith method was selected as the best method to use for determining crop water requirements as part of formulating a National Document of crop water requirements in Iran (Ministry of Agriculture, 1998). To arrive at this decision, consultations took place between the Ministry of Jihad-e-Agriculture (Agricultural Engineering Research Institute, AERI, and the Soil and Water Research Institute, SWRI), Deputy of Soil and Water of the Meteorological Organization, and other interested organizations, as well as contributing universities. These consultations occurred under the leadership and supervision of an expert committee.

To develop these national crop water requirements, a computer program was prepared for calculation of daily evapotranspiration (ET_0) based on the Penman-Monteith method. This program uses existing data from the Meteorological Organization, consisting of daily readings from multiple stations, to calculate ET_0 values and store them in another file. Since the number of statistical years varies for different weather stations, a statistical index period of 25 years (1970-1995) was developed and daily ET_0 for each station was calculated.

In this manner, a total of 9,125 different ET_0 values were calculated. For stations where the period of record was less than 25 years, statistical methods were used to standardize the existing data to the 25-year index period. A total of 363 climatic stations were utilized for this study. The Penman-Monteith method was applied to data from these 363 stations. After calculating for each statistical year, the average of 25 years of crop water requirements for each day of the year was computed. In order to evaluate the results of the Penman-Monteith method, ET_0 was calculated using nine other conventional methods. The results were compared with one another to facilitate understanding any wide differences.

There are 618 plains or agricultural areas in Iran that have been investigated and categorized for the Comprehensive Agricultural and Watershed Basins Project as part of the Comprehensive Water Project (JAMAB) in Iran. Considering the location of the meteorological stations of each valley, and taking into account the climatic conditions of each plain, representative stations for each valley were determined and water requirements of each valley were estimated and specified.

Determinations were individually made for all annual crops and fruit trees of each plain, and for each case the following information was specified and saved in the computer program: area under cultivation, planting and harvesting date and stages of various growth period, [found using the facilities of different deputies of the Ministry of Jihad-e-Agriculture and results of the study of comprehensive project for development of agriculture].

With the change in the concept of reference plant in the Penman-Monteith method, the plant coefficient has been modified accordingly. For inclusion of a plant factor during various stages of growth, a special computer program was developed. This program provides the curves for changes in plant coefficient factor during the growth period and its value for each day of growth. By drawing the modified curves, it makes available the necessary means for determining the required index during the growth stage. For particular plants such as date palm, citrus, mango, and banana, ambiguities were cleared up by examining research results and scientific papers, and by inviting specialists to the consultation meeting. The total irrigation requirement is reduced by any amount of effective rainfall (R_e) that is received.

Priorities for Allocation and Water Pricing for Optimum Use of Water

As already noted, the major consumer of water (more than 93 percent) in Iran is the agriculture sector. The average water productivity of agriculture (crops and orchards) is 0.7 kg per cubic meter. Increase in the economic value of water is one of the major objectives in the Economic Development Programs of Iran. Increase in the economic value of water will be possible when the yield or return per specific volume of water increases. For this reason it is preferable to use the available water supply for producing commodities with higher economical efficiency, or to use it in regions where it returns more economic value.

Based on the above discussions, determination of cropping pattern for each region, determination of crop water requirements, and finally volumetric allocation of water have been considered to be the important objectives for increasing the economic value of water. For optimum use of water allocated to the farmers, the following policies are considered:

- Control of water resources and volumetric allocation of water to the farms is based on crop water requirements and recommended irrigation efficiencies.
- Based on current law (established in 1983) price for regulated surface water is between 1-3 percent of the value of the cultivated crops.
- Based on the 1983 law, water pumping from groundwater resources must be in accordance with the crop water requirement and proposed cropping pattern in each region. In this case, price for groundwater resources is 0.25-1.0 percent of the commercial value of crop yield.
- Subsidies for water charges and supervision charges will be levied on farmers whose yields are higher than average.
- Water allocation will be terminated to the farmers who in two successive years consumed water at more than the permissible level.
- Policies will encourage the farmers to use less water and maintain their production at reasonable levels using proper management practices.

Cropping Pattern and Water Requirements for Different Regions

The cropping pattern in each soil and water resources development program is determined by considering factors such as climate, quantity and quality of soil water resources, social needs, livestock, food processing industries, government policy, farming culture of the region, job creation, marketing, transport, vertical and horizontal development of agricultural lands, crop rotation, mechanization, crop diseases, agricultural commodities, potential of executive organizations, health, environment, and sustainable development.

For setting this policy in each region the following programs are considered:

- Investment for rehabilitation of present irrigation networks.
- Control and prevention of over-extraction of water resources.

- Education as the basis for improvement of the methods. Efforts should be focused on education and extension to farmers for optimum use of water and increase in water productivity considering soil fertility, sustainable agriculture, and other environmental challenges.

APPROACHES

The challenges and opportunities for improving water productivity in the future are summarized in three system components as follows:

- biological (crop),
- environmental, and
- management.

However, it is well known that water productivity is the interaction and consequence of all foregoing components in any irrigation system. In the biological realm, drought resistant varieties of crops play an important role in improving water productivity. In this case, genetic improvement of the irrigated crops has been a part of the effort. Specific breeding programs have also been aimed at improving water productivity in irrigated and dryland agricultural systems, using both conventional breeding and genetic engineering. A primary issue in this regard is the study of the interactions between soil fertility, plant nutrition, and water management at the level of plant, plot, and system, and all the way up to the basin level.

Although parts of the objectives discussed above have been achieved, there are big gaps between the results obtained from research and the practical application of these results by the users. This is primarily because there has been little interaction between scientists, extension agents, and farmers for practical application of the latest findings. Although little is known about the factors causing the existing gaps, they are common in agricultural water management. The assumption is that new technologies have been picked up spontaneously and used in incorrect ways with insufficient attention by farmers.

There are many aspects involved in how farming communities integrate biophysical and socioeconomic issues to adapt their agricultural systems to optimize water use efficiency and water productivity. Many important issues involving anthropology and socioeconomic sociology should also be investigated. Specifically in water institutions study agendas the following concepts should be addressed:

1. Innovative institutions that would deal effectively with problems such as groundwater and waste water management.
2. Rules of prices and regulations for improving water management.

3. New methods for enhancing stakeholder participation in institutions and in defining water policies and water management.

4. Technical approaches for strengthening irrigation projects, such as equipping and renovating lands and irrigation networks (both in traditional and newly developed projects), and expansion of appropriate pressurized systems.

5. Agricultural approaches such as selection and operation of appropriate crop patterns in different regions.

In addition to the above-mentioned measures, the following technical approaches, management approaches, organizational approaches, and agricultural (biological) approaches are being performed for strengthening the laws that have been instituted by the government.

Technical Approaches

- Equipping and renovating lands, including land leveling, land consolidation, and drainage and land reclamation.
- Constructing new irrigation networks and lining traditional irrigation networks.
- Expanding pressurized irrigation systems, including enforcement of sprinkler irrigation systems for uniform application of water, enforcement of low energy precise application (LEPA) irrigation for decreasing evaporation and preventing wind effects, and enforcement of micro-irrigation (surface and subsurface) to decrease evaporation effects.

Management Approaches

- Making recommendations concerning irrigation programming.
- Enforcing water distribution systems using gated pipe in the field to improve conveyance efficiency and reduce deep percolation and evaporation.
- Supplying water requirements of the crops during critical growth stages.
- Reusing surface runoff and drainage water.
- Making use of marginal water.

Organizational Approaches

- Establishing a water utilization organization.
- Providing government subsidies for pressurized irrigation systems and other infrastructure.
- Extending applied research, and improving education and extension programs.
- Improving and applying suitable pressurized irrigation system technologies.

Agricultural (Biological) Approaches

- Selecting drought resistant crop varieties.
- Selecting appropriate crops.
- Optimizing use of other agricultural inputs, such as fertilizers, herbicides, and other supplements.
- Implementing a proper crop rotation program.

SELECTED BIBLIOGRAPHY

Allen, R., G.M. Smith, W.O. Pruitt, and L.S. Pereira. 1966. Modifications to the FAO crop coefficient approach. Proceedings of the ASAE International Conference on Evapotranspiration and Irrigation Scheduling, Nov. 3-6 San Antonio; TX.

Blaney, H.F. and W.D. Criddle. 1950. USDA Soil Conservation Service Tech. Paper No. 96, p. 48.

Doorenbos, J. and A.H. Kassam. 1979. Yield response to water. FAO Irrigation Drainage Paper 33, United Nations, Rome.

Doorenbos, J. and W.O. Pruitt. 1977. Guidelines for predicting crop water requirements. Irrigation and Drainage Paper 24, Rev., Rome. p.156.

Dehghani, S.H, A. Alizadeh, and A. Keshavarz. 1999. Implementation of water use pattern in terms of volumetric supply of water to farmers, IAERI.

Monteith, J.L. 1965. Evaporation and environment. Symp. Soc. Exp. Biol. 19, p. 205-234.

Ministry of Agriculture and Meteorological Organization. 1998. National documents of water; Crop water requirements, Vol. 30.

Penman, H.L. 1948. Natural evaporation from open water, bare soil and grass. Proc. Roy. Soc. A. 193:120-145.

Smith, M., R. Allen, J.L. Monteith, L.A. Pereira, and A. Segeren. 1991. Report on the expert consultation for the revision of FAO methodologies for crop water requirements. FAO/AGL, Rome.

Smith, M., R. Allen, and L. Pereira. 1996. Revised FAO metholoogy for crop water requirements. Proceedings of the ASAE International Conference on Evapotranspiration and Irrigation Scheduling. Nov. 3-6, San Antonio, TX. Pp. 16-123.

Thornthwaite, C.W. 1948. An approach towards a rational classification of climate. Geog. Rev. 38, p. 55-94.

MUNICIPAL WATER USE

St. Petersburg, Florida, Dual Water System: A Case Study

James Crook

INTRODUCTION

The city of St. Petersburg, Florida, is a largely residential peninsular community located on Florida's west-central coast. It is bound on the east and south by Tampa Bay and on the west by the Gulf of Mexico. St. Petersburg has a population of approximately 250,000. The Tampa Bay area receives an average of 140 centimeters (cm) (55 inches [in]) of rainfall annually, nearly half of which falls during the months of June, July, and August. Approximately 100 cm (40 in) of the 140 cm (55 in) are lost to evapotranspiration, leaving only 40 cm (15 in) available for potable and other uses. Due to the region's flat topography, there is little opportunity to impound water as a water supply source. Thus, while some of the rainfall percolates into the underground and enhances the groundwater supply, the majority of the rainfall remaining after evapotranspiration becomes runoff and eventually flows into the sea. The water supply problem is further compounded by a continuing influx of new residents to the area, many of whom choose to live in coastal areas where the groundwater supply is most limited because of seawater intrusion.

St. Petersburg has no significant surface water or groundwater suitable for potable water supplies within its corporate boundaries. As a result, water is obtained from adjacent counties from which several other municipal governments also obtain their water supplies. This situation, coupled with restrictive wastewater discharge requirements, led St. Petersburg to develop one of the largest urban water reuse systems in the world.

The initial portion of the retrofit system went into operation in 1977. Since that time it has grown both in volume of reclaimed water delivered and number of customers. The dual water system currently serves almost 10,000 customers throughout the city, including more than 9,400 residential customers for landscape irrigation. In 2001, 79,000 cubic meters per day (m^3/d) (21 million gallons per day [mgd]) of reclaimed water were used by system customers to irrigate more than 2,500 hectares (ha) (6,200 acres [ac]) of parks, school grounds, golf courses, and commercial and residential property. The reclaimed water also is used for cooling at a resource recovery facility and in air conditioning units at commercial buildings, including a large domed sports stadium.

WATER RESOURCES

By 1900, the municipal wells located in St. Petersburg were being pumped for increasingly longer intervals because of a growing population. By the mid-1920s, chloride in the groundwater began to increase due to seawater intrusion. Realizing that it was facing a potential water crisis, the city entered into a contract with a private company to provide St. Petersburg with a new water supply. The company purchased a section of land in adjacent Hillsborough County, developed a well field, constructed a water treatment plant, and laid approximately 48 kilometers (km) (30 miles [mi]) of 90-centimeter (36 in) water main from the water treatment facility to a water repumping station the company constructed north of the city. In the early 1940s, St. Petersburg purchased the company's assets, including a second undeveloped section of land in Hillsborough County, as well as Weeki Wachee Springs, located in Hernando County. The water company bought the spring with the intention of utilizing it as a water source at some future date when it would be more cost-effective. The spring is located more than 97 km (60 mi) from St. Petersburg. By the early 1960s, the unused property in Hillsborough County was developed as a groundwater source of supply for the city. In the late 1960s, a third property was purchased and developed as a well field. It was located approximately 64 km (40 mi) north of St. Petersburg in Pasco County.

When St. Petersburg joined with Pinellas County in the early 1970s to develop another well field in Pasco County, Pasco, Hillsborough, and Hernando Counties joined together to have legislation enacted to block any future water development by municipalities outside of their jurisdiction. The counties became alarmed that they might not be able to provide adequate water for their own growing populations because of St. Petersburg's water withdrawals.

St. Petersburg faced a two-fold problem in the early to mid-1970s. First, it needed additional water, but it was uncertain if permission could be obtained to develop a new supply. Due to costs, ecological concerns, and the possibility of worsening already strained relations with other counties, the development of Weeki Wachee Springs as a water supply source for the city was not considered a workable alternative. One option was to drastically reduce its future water demand. Secondly, because of rapid population growth, the city's four waste-

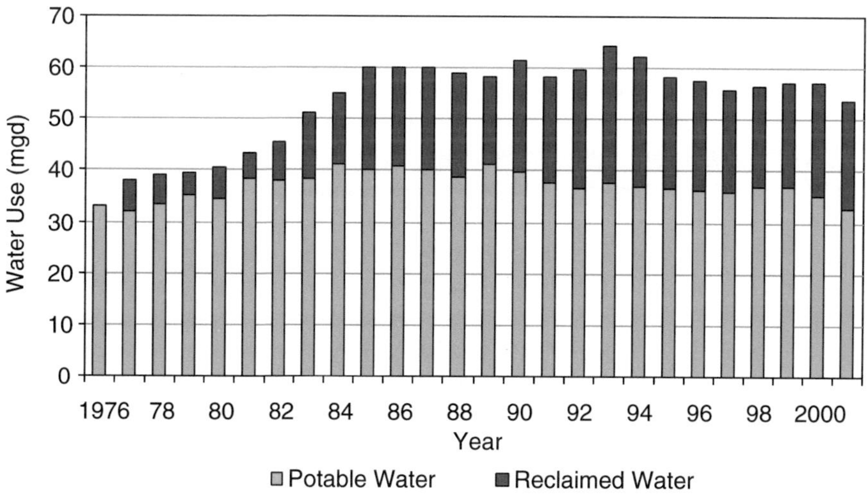

FIGURE 1 Water use in St. Petersburg, Florida.

water treatment plants needed to be enlarged. Concurrently, the Florida Legislature enacted a bill in 1972 that required all communities in the Tampa Bay area to cease discharging to Tampa Bay or to treat their wastewater with advanced wastewater treatment (AWT) processes to reduce biodegradable organic matter and nutrients to very low levels. Wastewater discharged to surface waters could not exceed 5 mg/l biochemical oxygen demand (BOD), 5 mg/l suspended solids, 3 mg/l total nitrogen, and 1 mg/l total phosphorus. The city evaluated the alternatives and, based on the cost of constructing and operating AWT facilities and considering the water supply problems, opted to upgrade the plants to tertiary treatment (i.e., secondary treatment, coagulation, filtration, and disinfection) and implement a water reuse and deep well injection program that would result in a zero discharge to surface waters.

In 2001, the reclaimed water system supplied 79,000 m^3/d (21 mgd) of the 204,000 m^3/d (54 mgd) total water provided by the city's Utility Department. Figure 1 indicates the growth of the reclaimed water system from its inception in 1977 to 2000. Because of the lowered demand for potable water, the necessity for a water plant expansion has been postponed and may not be needed at all if current water usage trends continue.

RECLAIMED WATER SYSTEM

St. Petersburg's reclaimed water system has several component parts, which are described in the following paragraphs.

Water Reclamation Plants

There are four water reclamation plants (WRPs) in St. Petersburg: Albert Whitted (Southeast) WRP; Northeast WRP; Northwest WRP; and Southwest WRP. They all have been upgraded to conform to the current Florida Department of Environmental Protection water reuse regulations. The plants have treatment capacities ranging from 47,000 m^3/d (12.4 mgd) to 76,000 m^3/d (20 mgd), with a total rated capacity of 260,000 m^3/d (68.4 mgd). The treatment process train is essentially the same at all of the four water reclamation plants and consists of grit removal, activated sludge biological treatment, secondary clarification, chemical coagulation, filtration, and disinfection. Covered storage of the reclaimed water is provided at each of the wastewater reclamation plants. The reclaimed water ground storage tanks have a total capacity of 95,000 m^3 (25 million gallons).

The initial reclaimed water distribution system was limited to serving irrigation water to golf courses, parks, school grounds, and large commercial areas. In 1981, the city applied for grant funding from the U.S. Environmental Protection Agency to expand the reclaimed water distribution system. A study conducted in support of the grant application identified four areas in the city where groundwater quality was deemed especially poor for irrigation. These areas were located adjacent to the coast and were designated "water quality critical" because the shallow groundwater supplies were either inadequate or contained high concentrations of chlorides or iron. Many of these locations were dredge and fill sites, where expensive waterfront homes were constructed. This study led to expansion of the reclaimed water system into residential areas. From 1977 through 1987, St. Petersburg spent more than $100 million upgrading and expanding the four water reclamation plants and constructing over 320 km (200 mi) of reclaimed water pipelines.

In 2001, the total average daily flow from the four water reclamation plants was approximately 160,000 m^3/d (42 mgd). The volume of reclaimed water used amounted to 42 percent of the total water supplied by the St. Petersburg Utilities Department, including about 5,500 m^3/d (1.5 mgd) used for in-plant purposes at the reclamation plants. The dual distribution system provided reclaimed water to almost 10,000 customers. Of these, more than 9,300 were individual residences receiving reclaimed water for lawn and ornamental plant irrigation, and 440 were commercial or industrial customers who received reclaimed water for a variety of applications, including irrigation and cooling water. The treated wastewater that was not reused was pumped down deep injection wells for disposal.

Injection Wells

Deep injection wells are used to dispose of excess reclaimed water and inadequately treated wastewater. The city operates a total of 10 injection wells at the four WRPs. There are either two or three wells located at, or adjacent to, each reclamation plant. The injection wells range in diameter from 50 cm (20 in) to 76 cm (30 in). The wells penetrate to a saltwater aquifer approximately 300 m (1,000 ft) below the land surface. The water in this aquifer contains approximately 22,000 mg/l of chlorides, precluding its use as a water supply. The total injection capacity is about 530,000 m^3/d (140 mgd).

Injection well testing indicated that the injection zone formation in St. Petersburg is sensitive to the amount of solids in the reclaimed water. Even though the wastewater is injected into a fractured and highly transmissive dolomite and limestone formation, slight pressure increases have been observed when suspended solids in the wastewater are greater than 10 mg/l. Acidization of the injection wells with concentrated hydrochloric acid restored the injection capacities of the affected wells.

Water quality monitoring via observation wells at the Southwest WRP indicated that the fresh, nonsaline reclaimed water is very buoyant in the saline injection zone. A large amount of mixing was observed in the injection plume as it moved out horizontally away from the injection wells. Even after several years of injection, the onsite injection zone observation wells still show a mixture of reclaimed water and saline water. It was hoped that the injected reclaimed water would form a bubble in the aquifer such that it could be stored in the underground and extracted as needed in the future. However, a significant reclaimed water lens has not been observed to form after prolonged injection.

Distribution System

As previously stated, the reclaimed water dual distribution system served almost 10,000 customers in 2001. Reclaimed water is delivered through more than 160 km (100 mi) of trunk and transmission mains ranging from 25 cm (10 in) to 120 cm (48 in) in diameter. Local service is provided through more than 300 km (190 mi) of small diameter distribution pipe ranging from 5 cm (2 in) to 20 cm (8 in) in diameter. The transmission mains from all four WRPs are interconnected so that reclaimed water flow and pressure can be maintained on the entire distribution network when any one plant is taken out of service. The reclaimed water system incorporates five city owned and operated booster pump stations and four privately owned and operated booster pump stations to provide reclaimed water for all of the applications throughout the city.

One of the early decisions made during development of the dual distribution system involved color-coding all polyvinyl chloride (PVC) pipes, using blue for potable water, green for sewers, and brown for reclaimed water. All buried ductile iron pipe was affixed with a coded brown tape to denote it as part of the reclaimed water distribution piping network. Hydrants installed on the system to flush the lines and serve as a backup to the fire protection system were also color-coded. Reclaimed water is not used as the primary fire protection source because it is considered to be an interruptible source. Reclaimed water valve boxes are square to differentiate them from potable water valve boxes, which are round. Several years ago purple became the standard color used to color-code reclaimed water pipes, valves, and appurtenances. Hence, St. Petersburg has switched to the purple color as the standard for identifying the reclaimed water system.

Nine monitoring wells are scattered throughout the city, because the entire community is considered to be a reclaimed water irrigation site. Most of the wells are located in the major irrigation areas, such as golf courses and school grounds. Water quality samples are routinely obtained at different locations on the distribution system and analyzed for fecal coliform organisms and chlorine residual. System pressure also is monitored at key locations.

Residential Irrigation

Residential landscape irrigation with reclaimed water is voluntary in St. Petersburg. Reclaimed water lines are brought into an area when at least 50 percent of the residents in that area petition for service and agree to connect to the reclaimed water system. The residents who hook up to the system pay the cost of extending distribution lines to serve them, which typically ranges from $500 to $1,200 per customer. The total connection charge for a 1.6-cm (5/8-in) or 1.9-cm (3/4-in) line is $295, consisting of a $180 tapping fee and $115 for a backflow prevention device on the potable water line. Reclaimed water costs $10.36 for the first 0.4 ha (1 ac) and $5.92 for each additional 0.4 ha (1 ac) or portion thereof.

All potable water services located in areas where reclaimed water service is available are protected with a cross-connection control device. Dual check valves are installed at each potable water meter. The backflow preventers are intended to protect the potable water system from possible illegal or inadvertent cross-connection of the reclaimed water system and potable water system.

A typical residence in St. Petersburg uses as much as 110 m^3 (30,000 gallons) per month of reclaimed water for landscape irrigation during peak demand periods, assuming a residential lot size of 650 m^2 (7,000 ft^2). The average irrigation rate is 4 cm/week (1.5 in/week). The average home discharges approximately 23 cm^3 (6,000 gallons) of sewage per month into the sanitary sewer system. Thus, it requires about five sanitary sewer customers in order to provide an adequate supply to one reclaimed water customer during peak demand periods. Reclaimed water reuse for residential irrigation in St. Petersburg is not metered, and surveys have shown that most residential customers use about 20 percent more reclaimed water than necessary for proper irrigation. Irrigation rates in excess of 4 cm/week (1.5 in/week) increase opportunistic weed infestations and the incidence of fungal diseases in many turf grass species.

Use of nutrient-rich reclaimed water has resulted in reduced fertilizer costs for the system's irrigation customers. Application of approximately 4 cm (1.5 in) of reclaimed water per week has been estimated to provide 50 percent of the nitrogen, phosphorus, and potassium requirements for horticultural and agricultural purposes.

WATER QUALITY AND HEALTH CONSIDERATIONS

The Florida Department of Environmental Protection has adopted comprehensive water reuse criteria, which were most recently revised in 1999. The criteria are summarized in Table 1. For residential and public access irrigation, the regulations require that wastewater receive secondary treatment, filtration, and disinfection such that the fecal coliform level is below detectable limits in 75 percent of the samples analyzed over a 30-day period and does not exceed 25 fecal coliforms/100 ml at any time. A minimum total chlorine residual of 1.0 mg/l is required after at least 15 minutes contact at peak hour flow. The regulations also specify a maximum BOD limit of 20 mg/l, a total suspended solids limit of 5 mg/l prior to disinfection, and continuous monitoring of turbidity and chlorine residual. The intent of the criteria is to assure that the treated water is essentially free of pathogens.

Several hundred reclaimed water samples have been analyzed for virus since implementation of the dual distribution system. Detectable levels of virus have occasionally been observed, but improvements in treatment reliability at the WRPs throughout the years have greatly reduced the number of samples having detectable levels of virus. The samples that were positive generally contained less than one enteric virus per 100 liters. It is noteworthy that there have not been any reported cases of illness or disease resulting from the use of reclaimed water at St. Petersburg.

SYSTEM PROBLEMS

In the early stages of the reclaimed water program, reclaimed water was stored in an open pond at one of the WRPs prior to pumping it to the distribution system. Unfortunately, the turnover rate of reclaimed water in the pond was low, and nutrients in the water promoted duckweed and algae blooms. In addition, the introduction into the pond of palm tree seeds dropped by birds, and other particulates, caused considerable clogging of irrigation spray nozzles. These problems were corrected by storing the finished reclaimed water in covered ground storage tanks at all four WRPs.

During the first year of operation of the residential reclaimed water system, it was discovered that backflow preventers on potable water lines were constraining the expansion of water within some residential internal water systems. Backflow preventers did not allow water to expand normally when the hot water temperature increased, and safety valves on hot water heaters allowed the excess water to be discharged into homes. Installing separate pressure relief valves on outside potable water hose bibbs at the residences solved the problem.

TABLE 1 Florida Treatment and Quality Criteria for Reclaimed Water

Type of Use	Water Quality Limits	Treatment Required
Restricted public access area irrigation,[a] industrial uses[b]	200 fecal coli/100 ml 20 mg/l TSS[c] 20 mg/l CBOD[d]	Secondary & disinfection
Public access area irrigation,[e] food crop irrigation,[f] toilet flushing,[g] fire protection, aesthetic purposes, dust control, commercial laundries, vehicle washing, other uses[h]	No detectable fecal coli/100 ml[i] 5.0 mg/l TSS 20 mg/l CBOD	Secondary, filtration, & disinfection
Rapid infiltration basins, absorption fields	200 fecal coli/100 ml 20 mg/l TSS 20 mg/l CBOD 12 mg/l NO_3 (as N)	Secondary & disinfection
Rapid infiltration basins in unfavorable geohydrologic conditions	No detectable fecal coli/100 ml[i] 5.0 mg/l TSS Primary & secondary drinking water standards	Secondary, filtration, & disinfection
Injection to groundwater	No detectable fecal coli/100 ml[i] 5.0 mg/l TSS Primary & secondary drinking water standards	Secondary, filtration, & disinfection
Injection to formations of Floridan or Biscayne Aquifers having TDS <500 mg/l [j]	No detectable fecal coli/100 ml[i] 5.0 mg/l TSS 3 mg/l TOC[k] 0.2 mg/l TOX[l] Primary & secondary drinking water standards	Secondary, filtration, disinfection, & activated carbon adsorption
Discharge to Class I surface waters (used for potable supply)	No detectable fecal coli/100 ml[i] 5 mg/l TSS 20 mg/l CBOD 10 mg/l NO_3 (as N) Primary & secondary drinking water standards	Secondary, filtration, & disinfection

continued

TABLE 1 Continued

SOURCE: Florida Department of Environmental Protection (1999).

[a]Sod farms, forests, pasture land, areas used to grow trees and fodder, fiber, and seed crops, or similar areas.

[b]Contact between reclaimed water and food or beverage products prohibited.

[c]TSS (total suspended solids) in reclaimed water used for subsurface irrigation systems cannot exceed 10 mg/l.

[d]CBOD = carbonaceous biochemical oxygen demand.

[e]Residential lawns, golf courses, cemeteries, parks, landscaped areas, highway medians, or similar areas.

[f]Direct contact between reclaimed water and tobacco and citrus is allowed, as is direct contact between reclaimed water and edible crops that are peeled, skinned, cooked, or thermally processed before consumption; also allowed for all edible crops where irrigation methods preclude direct contact between reclaimed water and crops.

[g]Only allowed where residents do not have access to plumbing system. Not allowed in single-family residences.

[h]Flushing of sanitary sewers and reclaimed water lines, mixing of cement, manufacture of ice for ice rinks, and cleaning roads, sidewalks, and outdoor areas.

[i]No detectable fecal coliform organisms/100 ml in at least 75 percent of the samples, with no single sample to exceed 25 fecal coliform organisms/100 ml.

[j]TDS = total dissolved solids.

[k]TOC = total organic carbon.

[l]TOX = total organic halogen.

In about 1985, St. Petersburg began receiving complaints from some residential homeowners claiming damage to ornamental plants and trees caused by irrigation with reclaimed water. Investigation revealed that chloride levels in the reclaimed water were as high as 700 mg/l at times. In response, the city conducted a research study and found that chloride levels above 400 mg/l in irrigation water for an extended time period damages salt-sensitive species of plants. A total of 205 common ornamental plant species was evaluated for tolerance to chloride; it was found that three types of plants (crape myrtle [*Lagerstroemia* spp.], azaleas [*Rhododendron* spp.], and Chinese privet [*Ligustrum sinense*]) have extremely low salt tolerances and should not be irrigated with reclaimed water. The problem was caused by infiltration of seawater into sewers near the coast and was solved by reducing seawater intrusion through an aggressive infiltration/inflow correction program, by mixing high chloride reclaimed water with reclaimed water containing low concentrations of chloride, and by diverting some reclaimed water containing very high chloride levels to the deep wells for disposal. Reclaimed water chloride levels are now kept below 400 mg/l, and complaints have ceased.

Although approximately 60 percent of the effluent is injected into deep wells for disposal on a yearly basis, there are times when the demand for reclaimed water can exceed the supply. Demands increase substantially during the hot, dry spring months when wastewater flows are at a minimum, and single day consecutive day peak distribution system demands stress the supply. The city addressed this problem by providing additional storage. Other measures currently being considered include: metering the reclaimed water to control overuse; restricting irrigation during critical periods; restricting further expansion of the reclaimed water system; developing an aquifer storage and recovery system to seasonally store reclaimed water and recover it for use during high demand periods; and developing informational programs to further educate the public about proper use techniques and lawn management.

WATER CONSERVATION EFFORTS

While St. Petersburg's reclaimed water system is the cornerstone of the city's water conservation program, a variety of conservation efforts has been implemented to reduce potable water consumption. In addition to reclaimed water, the water conservation efforts include operational, regulatory, economic, and education outreach programs.

Operational Programs

In 1995, an indoor retrofit program was initiated that included providing water conserving showerheads, toilet tank banks, faucet aerators, and leak detection tablets. Retrofit kits were distributed to more than 145,000 customers, resulting in a savings of 4,500 m^3/d (1.2 mgd) of potable water. A toilet replacement program was begun in 1997, offering financial incentives to replace existing toilets with low flush fixtures. To date, more than 15,000 toilets have been replaced, resulting in a water savings of 1,900 m^3/d (0.5 mgd).

Unaccounted water loss is another focus of the water conservation program. Aged meters become less reliable and frequently under-record water usage. A meter replacement program was initiated to replace old and inaccurate meters, resulting in customers now paying for all water used. Each year more than 9,300 meters are taken out of service and replaced. In addition, a program was established to minimize water loss due to leakage. More than $500,000 per year is allocated for repairs and leak detection within the water distribution system. As a result, unaccounted water volume has been reduced from 12 percent of the total water production to 5 percent.

Regulatory Program

In 1994 the Southwest Florida Water Management District (SWFWMD) declared a water shortage in several counties, including Pinellas County (which includes St. Petersburg). In response, St. Petersburg adopted water restrictions for irrigation of lawns and landscapes. Additional restrictions imposed by SWFWMD further limited irrigation with potable water and well water, e.g., irrigation was restricted to one day per week. The city established an enforcement program, which has recorded more than 2,000 outdoor water use violations in the last year alone.

Economic Incentives

An inverted utility rate structure was adopted by the city in 1985 to encourage water conservation. The water volume usage charges, as of January 2000, are shown in Table 2. St. Petersburg's wastewater volume charge is $0.64/m³ ($2.44/1000 gallons).

Educational Outreach

Adult educational programs include participation in public forums concerning water issues; provision of water conservation books and videos at libraries; weekly taped television broadcasts; booklets on residential xeriscape methods; online water conservation information via a website; annual public recognition awards; and fairs, festivals, and other community events promoting water conservation. In 2000, a youth education position was created to provide water conservation education through public schools and youth resource agencies. Presentations are made at schools, and educational materials are distributed to students. An annual Drop Savers Conservation Poster Contest is held every year; more than 10,000 children participated in the contest in 2000.

TABLE 2 Water Volume Charges for Single-Family and Multi-Family Residences and Commercial Customers

Amounts		1999 Rate		2000 Rate	
Residential	Commercial	$/m³	$/1000 gal	$/m³	$/1000 gal
First 21 m³ (5600 gal)	Up to average	0.39	1.49	0.42	1.61
Next 9 m³ (2400 gal)	Avg. to 1.4 times avg.	0.49	1.87	0.53	2.02
Next 27 m³ (7000 gal)	1.4 to 1.8 times avg.	0.67	2.53	0.72	2.73
Over 57 m³ (15,000 gal)	Over 1.8 times avg.	0.96	3.36	0.96	3.63

SUMMARY

The city of St. Petersburg operates one of the largest urban dual water systems of its kind in the world. The extensive use of reclaimed water has stabilized the use of potable water in St. Petersburg and eliminated the need to develop additional water supply sources in the near future. Several design and operational problems have been encountered and solved in this pioneering operation as the reclaimed water distribution system has expanded throughout the years. The reclaimed water has been shown to be safe and acceptable for the intended uses.

REFERENCE

Florida Department of Environmental Protection. 1999. Reuse of Reclaimed Water and Land Application. Chapter 62-610, Florida Administrative Code. Tallahassee: Florida Department of Environmental Protection.

Monterey County Water Recycling Projects: A Case Study

James Crook and Robert S. Jaques

INTRODUCTION

The Monterey Regional Water Pollution Control Agency (MRWPCA) began facilities planning to provide wastewater management services to northern Monterey County, California, in 1975. At that time, water reuse was considered to be an important element in the planning process. The Salinas Valley is an agricultural region in northern Monterey County where a wide variety of market crops are grown. Heavy agricultural and municipal groundwater demands beginning in the 1940s led to the development of severe groundwater overdrafting of the underlying aquifers, resulting in seawater intrusion from adjacent Monterey Bay. The intrusion front was advancing inland at a rate of approximately 150 meters/year (m/yr) (500 feet/year [ft/yr]). High salt levels in groundwater caused wells near the coast to be abandoned, and agricultural water supply wells and some community drinking water wells were threatened. This was a major factor in the decision to develop a regional wastewater management plan to provide reclaimed water for food crop irrigation in the Salinas Valley. By using reclaimed water for irrigation, growers could discontinue pumping from their wells, thus alleviating overdrafting of the groundwater. The wastewater management plan included eliminating nine older wastewater treatment plants by constructing a single centralized regional treatment facility. Figure 1 depicts the water reuse scheme.

A seven-year agricultural reuse demonstration study conducted at Castroville, California, initiated in 1976 and completed in 1987, determined that filtered, secondary effluent meeting a total coliform limit of 2.2/100 milliliters (ml) was acceptable for the spray irrigation of food crops eaten raw. During the study no

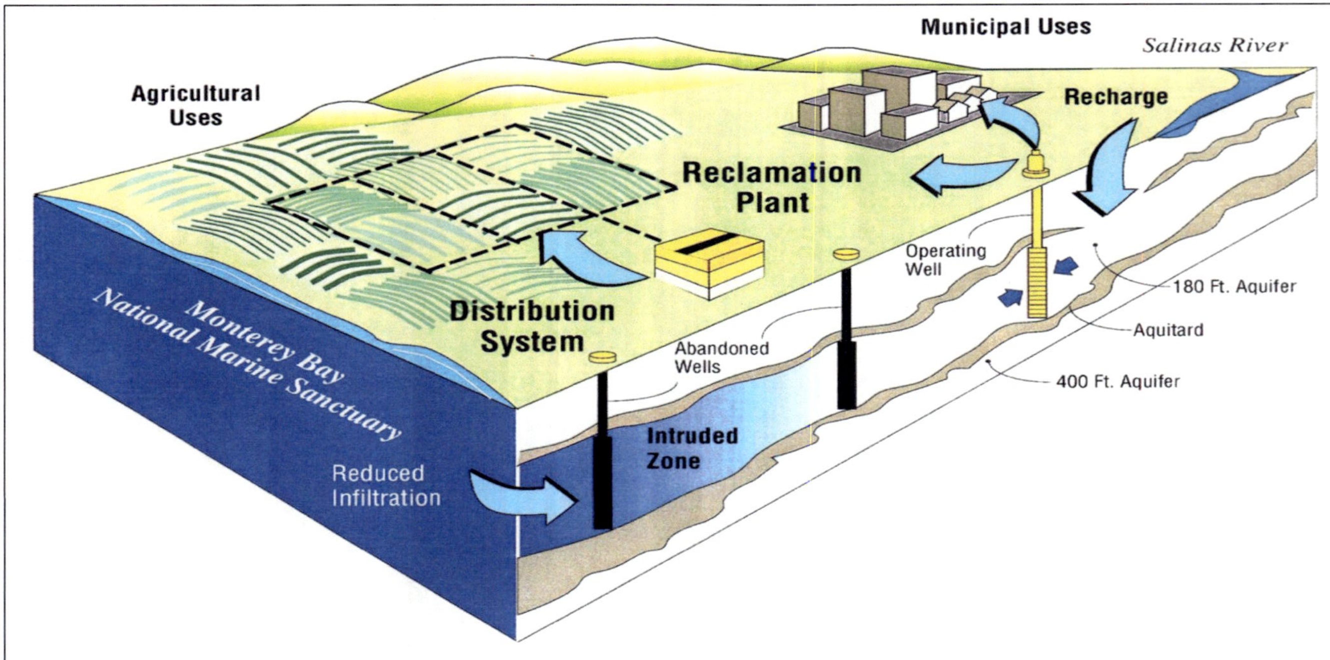

FIGURE 1 Schematic of the water reuse concept.

pathogenic organisms were detected in the reclaimed water, and spray irrigation with reclaimed water did not adversely affect soil permeability, did not result in heavy metal accumulation in the soil or plant tissue, and did not adversely affect crop yield, quality, or shelf life.

The 114,000 cubic meters per day (m^3/d) (30 million gallons per day [mgd]) regional wastewater reclamation facility was constructed adjacent to a regional secondary treatment plant to provide tertiary treated reclaimed water for agricultural applications. A distribution system to serve 4,800 hectares (ha) (12,000 acres [ac]) was constructed to deliver the reclaimed water. The regional wastewater reclamation facility began delivering 76,000 m^3/d (20 mgd) of reclaimed water for food crop irrigation in 1998. Reclaimed water is used to irrigate lettuce, celery, broccoli, cauliflower, artichokes, and strawberries. The water reclamation facility and distribution system are collectively known as the Monterey County Water Recycling Projects (MCWRP).

MONTEREY WASTEWATER RECLAMATION STUDY FOR AGRICULTURE

The Monterey Wastewater Reclamation Study for Agriculture (MWRSA), initiated in 1976 and completed in 1987, was an important step in the planning process for the MCWRP. A Task Force composed of representatives from federal, state, and local governments, the academic community, farm advisors, and local growers provided guidance in the planning and conduct of the study. The California State Water Resources Control Board and the U.S. Environmental Protection Agency (EPA) provided funding for the study, which cost a total of $7 million. The goal of MWRSA was to assess the safety and feasibility of agricultural irrigation using reclaimed water to irrigate vegetable crops that may be eaten raw. It included a five-year demonstration project.

STUDY DESCRIPTION

The tertiary treatment plant used secondary effluent from the existing Castroville Wastewater Treatment Plant as influent to the pilot tertiary treatment plant, which had two parallel treatment process trains as follows:

1. Up to 57 m^3/d (0.015 mgd) was treated using the treatment train specified in California's Wastewater Reclamation Criteria for reclaimed water used to irrigate food crops that could be consumed raw (secondary treatment, chemical coagulation, clarification, filtration, and disinfection); and
2. Up to 890 m^3/d (0.24 mgd) was treated using the direct filtration process (secondary treatment, low dose coagulant/polymer addition, filtration, and disinfection), which provided less extensive treatment than that specified in the California water reuse criteria.

In order to conform to California's water reuse criteria, both treatment trains were required to have a chlorine residual of at least 5 milligrams per liter (mg/l) after a minimum chlorine contact time of 90 minutes. In addition, the turbidity of the wastewater prior to disinfection was required to be 2 nephelometric turbidity units (NTU) or less. The product water was dechlorinated to avoid chloride burn of the vegetable leaves for the first three years of the field studies. Dechlorination was discontinued for the remaining two years with no detectable adverse impacts to the crops. The current California Department of Health Services Water Recycling Criteria for irrigation and other nonpotable uses are summarized in Table 1.

TABLE 1 California Water Recycling Criteria: Treatment and Quality Requirements for Nonpotable Uses of Reclaimed Water

Type of Use	Total Coliform Limits[a]	Treatment Required
Irrigation of fodder, fiber, & seed crops, orchards[b] and vineyards,[b] processed food crops, nonfood-bearing trees, ornamental nursery stock,[c] and sod farms;[c] flushing sanitary sewers	• None required	• Secondary
Irrigation of pasture for milking animals, landscape areas,[d] ornamental nursery stock, and sod farms where public access is not restricted; landscape impoundments; industrial or commercial cooling water where no mist is created; nonstructural fire fighting; industrial boiler feed; soil compaction; dust control; cleaning roads, sidewalks, and outdoor areas	• ≤23/100 ml • ≤240/100 ml in more than one sample in any 30-day period	• Secondary • Disinfection
Irrigation of food crops;[b] restricted recreational impoundments; fish hatcheries	• ≤2.2/100 ml • ≤23/100 ml in more than one sample in any 30-day period	• Secondary • Disinfection
Irrigation of food crops[e] and open access landscape areas;[f] toilet and urinal flushing; industrial process water; decorative fountains; commercial laundries and car washes; snow-making; structural fire fighting; industrial or commercial cooling where mist is created	• ≤2.2/100 ml • ≤23/100 ml in more than one sample in any 30-day period • 240/100 ml (maximum)	• Secondary • Coagulation[g] • Filtration[h] • Disinfection

continued

TABLE 1 Continued

Type of Use	Total Coliform Limits[a]	Treatment Required
Nonrestricted recreational impoundments	• ≤2.2/100 ml • ≤23/100 ml in more than one sample in any 30-day period • 240/100 ml (maximum)	• Secondary • Coagulation • Clarification[i] • Filtration[h] • Disinfection

[a]Based on running 7-day median.
[b]No contact between reclaimed water and edible portion of crop.
[c]No irrigation for at least 14 days prior to harvesting, sale, or allowing public access.
[d]Cemeteries, freeway landscaping, restricted access golf courses, and other controlled access areas.
[e]Contact between reclaimed water and edible portion of crop; includes edible root crops.
[f]Parks, playgrounds, schoolyards, residential landscaping, unrestricted access golf courses, and other uncontrolled access irrigation areas.
[g]Not required if the turbidity of the influent to the filters is continuously measured, does not exceed 5 nephelometric turbidity units (NTU) for more than 15 minutes and never exceeds 10 NTU, and there is capability to automatically activate chemical addition or divert the wastewater if the filter influent turbidity exceeds 5 NTU for more than 15 minutes.
[h]The turbidity after filtration through filter media cannot exceed an average of 2 NTU within any 24-hour period, 5 NTU more than 5 percent of the time within a 24-hour period, and 10 NTU at any time. The turbidity after filtration through a membrane process cannot exceed 0.2 NTU more than 5 percent of the time within any 24-hour period and 0.5 NTU at any time.
[i]Not required if reclaimed water is monitored for enteric viruses, *Giardia*, and *Cryptosporidium*.

SOURCE: Adapted from State of California (2000).

Three types of water were used during the field study: tertiary treated reclaimed water receiving the full treatment specified in the California water reuse criteria; tertiary treated reclaimed water receiving direct filtration; and local well water as a control. A 12-ha (30-ac) field site adjacent to the Castroville Wastewater Treatment Plant was divided into demonstration and experimental fields. Two 5-ha (12-ac) plots in the demonstration field were dedicated to irrigation using the direct filtration flow stream. Artichokes were grown on one plot and a succession of broccoli, cauliflower, lettuce, and celery was grown on the other plot. In the experimental field, artichokes were grown continuously from 1980 to 1985, and row crops were planted in rotation during the same time frame. A split plot design in the experimental fields allowed evaluation of water type and fertilization rate. Multiple replicates of each water type and each fertilization rate were evaluated in the experimental field, resulting in a total of 96 plots occupying 1.2 ha (3 ac). Separate irrigation systems were constructed to supply the three water types to the plots. Local farming practices were followed in both of the fields throughout the study.

Several baseline studies were carried out prior to the start of the five-year field demonstration phase of the MWRSA study to ascertain the uniformity of the soil on the site of the experimental plots and to assure the safety of downwind areas from windblown aerosols during irrigation with reclaimed water.

The sampling program included the following:

- Sampling for metals and other chemicals included composite samples of each water type taken over a 3-day to 5-day period during each irrigation event.
- Grab samples were taken for other analyses, e.g., microorganisms and biochemical oxygen demand (BOD).
- Tail water resulting from furrow irrigation of row crops was analyzed for ten metals and sixteen chemical parameters.
- Surface soil samples were collected for bacteriological analyses within two days of irrigation, and soil profile samples were collected and analyzed for a variety of metal, chemical, and physical parameters. Samples also were collected and analyzed for cation exchange capacity, boron, and pH and salt content.
- Laboratory and field permeability tests were conducted. Edible and residual plant tissues were sampled and analyzed for bacteria, parasites, and metals, and edible portions of crops also were collected for metals analyses at each major harvest.
- Crop residues were sampled and analyzed for cadmium, zinc, and boron. Samples of edible tissue were taken from neighboring fields for bacteriological and metal analyses to provide comparative data for the study.
- The Castroville Wastewater Treatment Plant secondary effluent, tertiary treated effluent, well water used for irrigation, plant tissues, and soils were sampled for enteric viruses. The native virus concentration in the secondary effluent was low; therefore, virus-seeding studies using vaccine-grade poliovirus were carried out to estimate the virus removal efficiencies of the two pilot plants producing tertiary effluent.
- Groundwater in the demonstration fields was sampled for nitrate and other constituents.

In addition to the above, climatic conditions were continuously measured and recorded in order to aid in the evaluation of crop development and analyses, and a field study was performed to compare aerosols generated by spray irrigation with reclaimed water and well water.

MWRSA STUDY RESULTS

Tertiary Treatment: Of the two tertiary treatment trains evaluated during the study, the more extensive train (i.e., secondary treatment, chemical coagulation, clarification, filtration, and disinfection) was somewhat more effective in removing or inactivating seeded virus than the direct filtration treatment train (i.e., secondary treatment, low dose coagulant/polymer addition, filtration, and disinfection), although both treatment trains reliably removed or inactivated more than 5 logs of virus during seeding experiments. Naturally occurring viruses were not detected in the product water of either tertiary treatment pilot plant.

Irrigation Water Quality: Both types of reclaimed water had higher levels of most chemicals, including metals, than the native local groundwater. Electrical conductivity, total dissolved solids (TDS), boron, chloride, and sodium in the reclaimed waters were similar to each other but higher than levels in well water. The TDS levels in all three types of water were below the severe problem range for irrigation water, and the sodium adsorption ratio in all three waters was in the favorable range for irrigation water.

Virus Removal/Inactivation: Measurable levels of viruses were detected in 53 of 67 samples (80 percent) of secondary effluent. No naturally occurring viruses were detected in disinfected tertiary effluent from either pilot treatment train throughout the study, and no viruses were detected in any of the crop or soil samples. An environmental chamber was constructed in a laboratory to determine virus survival under field conditions. The laboratory study indicated that the time required for 99 percent inactivation of seeded viruses ranged from 7.8 to 15.1 days, depending on the type of crop. Subsequent field studies produced similar results, and no seeded viruses were detected in any soil samples after 12 to 14 days of exposure.

Bacteria and Parasites: Coliform organisms were occasionally found in all three types of irrigation water. None of the samples taken from the three water sources or the soil indicated the presence of *Salmonella*, *Shigella*, *Ascaris lumbricoides*, *Entamoeba histolytica*, or other parasites. Parasites were detected in plant tissues during the first year of the study, but there were no differences between the levels in reclaimed and well water.

Aerosols: Aerosol tests were conducted during both daytime and nighttime irrigation. Microorganism transport via aerosols generated during spray irrigation of tertiary effluent was not significantly different from transport via aerosols generated during spray irrigation with well water, thus indicating that aerosol transmission of bacteria originating in the reclaimed water was unlikely.

Heavy Metals: There was no significant difference in any of the nine heavy metals studied (cadmium, chromium, cobalt, copper, iron, lead, manganese, nickel, and zinc) among plots irrigated with the different water types. Except for copper, the concentration of the metals did not show an increase in the soil during the five-year study period. Copper did increase gradually for all water types but remained below the average for California soils during the course of the study. Heavy metal input from commercial fertilizer impurities was far greater than from irrigation waters and accounted for the differences observed in soil samples throughout the five-year study period. Analyses of edible plant tissues indicated no consistent significant differences in heavy metal concentrations between plants irrigated with reclaimed water and well water. Heavy metal concentrations in the three types of irrigation waters are shown in Table 2.

Crop Yield and Quality: Crop yield for most of the vegetables grown during the study was slightly higher for crops irrigated with either of the two reclaimed waters than with well water. It also was observed that increases in yield tended to level off at fertilizer applications below the commonly applied fertilizer application rate for the area, indicating that the typical full fertilization rate may be in excess of the crops' requirements. Field crop quality assessments, shelf life measurements, and visual inspection did not reveal any difference between produce irrigated with reclaimed water and produce irrigated with well water. The use of reclaimed water to irrigate produce did not result in any increased spoilage over that encountered for irrigation with well water.

Marketability: A marketing firm was commissioned to conduct a study to determine the key issues associated with marketability of crops irrigated with reclaimed water. Interviews were conducted with individuals involved with produce distribution, such as wholesale-retail buyers, brokers, and store managers. Responses indicated that produce grown in reclaimed water would be accepted, labeling would not be necessary, and factual information would be useful to respond to customer inquiries that may arise. The major requirement of buyers was for produce to have a healthy appearance and be aesthetically attractive.

TABLE 2 Heavy Metal Concentrations (in mg/l) in Irrigation Waters

	Well Water		Tertiary Effluent[a]		Tertiary Effluent[b]		
Heavy Metal	Range	Median	Range	Median	Range	Median	Irrigation Water Criteria[c]
Cadmium	ND[d]-0.1	ND	ND-0.1	ND	ND-0.1	ND	0.010
Zinc	ND-0.6	0.02	0.07-6.2	0.33	ND-2.08	0.195	2.0
Iron	ND-0.66	0.1	ND-2.3	0.05	ND-0.25	0.06	5.0
Manganese	ND-0.07	ND	ND-0.11	0.05	ND-0.11	0.05	0.20
Copper	ND-0.05	0.02	ND-0.05	ND	ND-0.04	ND	0.20
Nickel	0.001-0.2	0.04	0.002-0.18	0.04	0.004-0.2	0.04	0.20
Cobalt	ND-0.057	ND	0.001-0.062	0.002	ND-0.115	0.05	0.050
Chromium	ND-0.055	ND	ND	ND	ND	ND	0.10
Lead	ND	ND	ND	ND	0.001-0.7	0.023	5.0

[a]Full treatment: oxidation, coagulation, clarification, filtration, and disinfection.
[b]Direct filtration: oxidation, coagulation, filtration, and disinfection.
[c]From: U.S. Environmental Protection Agency (1972).
[d]ND = Not detected.

Adapted from Sheikh et al. (1990).

Respondents to the market study recommended that response to rumors that might occur regarding produce irrigated with reclaimed water should include clear, government-endorsed fact sheets and that support be given to developing an educational information program on the use of reclaimed water for crop irrigation.

Field Worker Health: The health status of each person assigned a field task during the field study was monitored through frequent questionnaires and initial and exit medical examinations. Neither questionnaire data nor medical examinations indicated any adverse health effects associated with working in fields irrigated with tertiary treated reclaimed water.

FULL-SCALE PROJECT

Based on the favorable results of the MWRSA study, a decision was made to design and construct a full-scale facility. Design of the treatment plant facilities, called the Salinas Valley Reclamation Project (SVRP), was completed in 1994 and was followed by design of the distribution system, which is known as the Castroville Seawater Intrusion Project (CSIP).

The SVRP includes the following facilities (design and regulatory criteria are summarized in Table 3):

- facilities to pump effluent from the existing secondary treatment plant to the new reclamation facility,
- rapid mixing of coagulant (alum) and flocculent chemicals (polymer or powered activated carbon),
- flocculation,
- dual media gravity filtration,
- disinfection using gaseous chlorine,
- diurnal flow equalization storage.

The CSIP distributes reclaimed water to 222 parcels of farmland within the 4,800 ha (12,000 ac) service area and includes the following:

- 74 kilometers (km) (46 miles [mi]) of reclaimed water transmission and distribution pipelines ranging in diameter from 0.2 meters (m) to 1.3 m (8 inches [in] to 51 in),
- 22 supplemental wells to augment reclaimed water flows at times of peak demand,
- 111 flow-metered turnouts for connection of irrigation piping by farmers,
- pressure, conductivity, and flow monitoring stations,
- a centralized control system,
- three booster pump stations,
- cathodic protection for ferrous metal piping.

TABLE 3 Treatment Process Design Criteria and Effluent Quality Requirements

Item	Value
Design flows	
Average daily flow (mgd)	29.6
Average daily flow (m^3/s)	1.30
Peak instantaneous flow (mgd)	38.5
Peak instantaneous flow (m^3/s)	1.69
Coagulation and flocculation	
Number of flocculation basins	2
Number of mixing components per basin	3
Alum dosage (mg/l)	2 to 15
Polymer dosage (mg/l)	0 to 0.18
Flocculator detention time @ average daily flow (minutes)	12
Filtration	
Type	Dual media gravity
Number of filters	6
Media depth	
Anthracite	1.22 m (4 ft)
Sand	0.30 m (1 ft)
Loading @ peak flow (liter/m^2/min)[a]	204
Loading @ peak flow (gpm/ft^2)[ab]	5
Average effluent turbidity (NTU)[a]	2
Maximum effluent turbidity (NTU)[a]	$\leq$ 5 95% of the time
Surface area	580 m^2 (6,240 ft^2)
Disinfection (by chlorination)	
Target combined chlorine residual (mg/l)[a]	5.0
7-Day median total coliform concentration (MPN/100 ml)[a]	2.2
Single maximum total coliform concentration (MPN/100 ml)[a]	23
Minimum length to width and length to depth ratios[a]	40:1
Minimum modal contact time (minutes)[a]	90

[a]Effluent quality requirement.
[b]Gallons per minute (gpm) per square foot (ft^2).

SOURCE: Jaques (1997).

FOOD SAFETY STUDY

Prior to startup an independent laboratory was hired to conduct a Reclaimed Water Food Safety Study, which continued after startup of the full-scale project. The primary objective of the study was to determine if any viable pathogenic organisms of concern to food safety, such as *E. coli* 0157:H7, *Cyclospora*, and *Salmonella* were present in reclaimed water produced by the SVRP. A secondary objective was to assess the ability of the treatment processes to remove or inactivate pathogens that might be present in the influent wastewater. Samples were collected and analyzed during the 1997 to 1999 timeframe.

The study did not detect any *E. coli* 0157:H7, *Cyclospora*, *Salmonella*, helminth ova, viable *Giardia*, or culturable natural (in situ) viruses. Only an extremely low number of *Cryptosporidium* (in only two instances) was detected in any of the tertiary treated reclaimed water samples. Some of the study results are presented in Table 4.

TABLE 4 Microbial Water Quality at Salinas Valley Reclamation Project

Organism	Raw Wastewater	Secondary Effluent	Tertiary Reclaimed Water
E. Coli 0157:H7 (MPN/100 ml)[a]	ND[b]	ND	ND
Fecal Coliform (MPN/100 ml)	$7 \times 10^6 - 30 \times 10^6$	$230 \times 10^3 - 800 \times 10^3$	ND
Legionella (colonies/liter)	ND	ND	ND
Salmonella (MPN/100 ml)	ND – 16	2.2 – 9.2	ND
Helminth (*Ascaris*) Ova/liter	NS[c]	NS	ND
Shigella (MPN/100 ml)	NS	NS	ND
Cyclospora (No./liter)	ND – 330	ND	ND
Giardia (cysts with internal structure/liter)	2,000 – 22,400	0.4 – 12.2	ND – 0.3[d]
Cryptosporidium (Oocysts with internal structure/liter)	ND – 200	ND – 1.8	ND – 0.41
Culturable Virus (MPN/liter)	NS	NS	ND

[a]MPN = most probable number.
[b]ND = none detected.
[c]NS = not sampled.
[d]All of the *Giardia* cysts detected in the tertiary reclaimed water were empty cysts and were therefore classified as nonviable.

SOURCE: Adapted from Jaques et al. (1999).

SUMMARY

The total capital cost of the Monterey County Water Recycling Projects was approximately \$78 million. The total cost to treat and deliver reclaimed water to agricultural areas is estimated to be approximately \$0.19/m^3 (\$225/acre-foot or \$0.90/1,000 gallon) excluding secondary treatment costs but including both debt service from low interest loans and operation and maintenance costs for the two components (i.e., treatment facilities and distribution network) of the MCWRP. The 114,000-m^3/d (30-mgd) wastewater reclamation plant was completed in 1997 and began delivering 76,000 m^3/d (20 mgd) of reclaimed water for food crop irrigation in 1998. Reclaimed water is used to irrigate lettuce, celery, broccoli, cauliflower, artichokes, and strawberries. The service area is about 4,800 ha (12,000 ac).

SELECTED BIBLIOGRAPHY

Jaques, R.S. 1997. Twenty years in the making—now a reality: Nation's largest project to provide recycled water for irrigation of vegetable crops begins operation. Proceedings of WEFTEC '97 (October 18-22) 6:143-152.

Sheikh, B., R.P. Cort, W.R. Kirkpatrick, R.S. Jaques, and T. Asano. 1990. Monterey wastewater reclamation study for agriculture. Res. Jour. WPCF 62(3):216-226.

State of California. 2000. Water Recycling Criteria. California Code of Regulations, Title 22, Division 4, Chapter 3. Sacramento: California Department of Health Services.

U.S. Environmental Protection Agency. 1972. Water Quality Criteria 1972. EPA/R3/73/033. Washington, D.C.

Identifying Microbial and Chemical Contaminants for Regulatory Purposes: Lessons Learned in the United States

Rebecca T. Parkin

SUMMARY

Identification of potentially hazardous contaminants is the first step in risk management paradigms and in complying with regulatory mandates. While efficient identification is a key to the development of the United States Environmental Protection Agency's Contaminant Candidate List for drinking water regulation, there is a growing interest in proactive identification schemes outside of regulatory frameworks. This interest is resulting in the development and application of methods previously not used for hazard identification. The purposes of this paper are to describe the methods used in the United States to identify drinking water contaminants for regulatory purposes, evaluate the strengths and weaknesses of those methods, list the lessons learned, describe newly proposed methods that are currently being developed for evaluation, and propose potential international collaboration.

INTRODUCTION

The vision of a world with healthy water for all has been an elusive but continuously important ideal for many centuries. An international coalition declared "all water users must be supplied with water that will protect them from any health risk" (Bourbigot, 2000). Despite significant advances in the twentieth century, many challenges remain before the world can achieve this ambitious goal. The World Health Organization's work plan for water sanitation includes numerous activities designed to protect human health through improved water management (WHO, 2002). Ways to identify and respond to drinking water contaminants must become more proactive so that a broader range of feasible response options can be developed.

In the United States, regulation of drinking water contaminants has increased rapidly in the past decade since 1990. However, enforceable regulations were first instituted in 1914 to reduce the spread of disease across state lines. Under the Interstate Quarantine Act of 1893, municipalities that used their water in interstate carriers (e.g., buses, ships, and trains) were required to meet a federal regulation for fecal coliform bacteria. In 1925, 1942, 1946, and 1962, the federal Public Health Service added more contaminants to this regulation. Evidence from the late 1960s and early 1970s indicated that many previously undetected contaminants were in public water supplies and posed potential human health effects. Following the publication of these environmental data and *Silent Spring* (Carson, 1962), public awareness and concerns increased. Although most states adopted the earlier federal standards for the regulation of local water supplies, federally mandated regulations for large-scale water supplies did not exist until 1974.

Before the Safe Drinking Water Act (SDWA) was passed in 1974, most drinking water contaminants were controlled through guidelines and non-enforceable mechanisms. A major goal of the SDWA was to protect public health by ensuring that contaminants in public water systems met national standards. Environmental surveys following passage of the SDWA revealed the presence of additional contaminants. As a result, in 1986 Congress required that 25 more substances (chemicals and pathogens) be regulated every three years. This requirement proved to be unachievable, especially without new resources. The lessons learned from the efforts to meet these requirements led to a Congressional mandate to regulate contaminants on the basis of occurrence in water, level of risk posed, and the probability of adverse health effects (NRC, 1997).

For most of the twentieth century, the United States identified hazardous substances and microbial pathogens in drinking water through investigations of disease outbreaks. Although this approach has been recognized as limited (Balbus and Embrey, 2002), it has been the traditional method used by epidemiologists. In 1970, the U.S. Environmental Protection Agency (EPA) was charged with controlling hazardous substances in drinking water and other media (United States Code Annotated, 1970). In response, the agency began developing procedures to identify and prioritize chemical contaminants. A standardized chemical risk assessment framework published in 1983 was used on a substance-specific basis (NRC, 1983). However, this process proved to be too resource intensive and time-consuming to prioritize the increasing number of contaminants needed for research and regulatory purposes.

In the mid-1980s, EPA reported that the classification of substances into categories based on the weight of scientific evidence was an important part of the process of setting public health priorities and began developing more efficient ways to do so (EPA, 1987). In 1996, Congress directed the EPA to develop a list of unregulated contaminants, known as the Contaminant Candidate List (CCL), every five years and to evaluate at least five contaminants every five years for potential regulation (SDWA, 1996). As evidenced by the process used to develop and publish the first CCL (CCL1) in 1998, this requirement proved to be very difficult to meet (NRC, 2001). Although EPA developed a risk-based conceptual model, the agency's planned process, including the publication of a Contaminant Identification Method incorporating public input, could not be implemented in time to meet the legislative deadline. Consequently, EPA staff compiled existing lists of contaminants from the Agency's programs (e.g., water, hazardous waste sites), obtained some external input, pared the list based on drinking water occurrence, and finalized the list with 50 chemicals and 10 microorganisms. One of the problems with this approach was that available occurrence data became the critical criterion for selection; potential health effects became secondary and were not considered for any substance deleted from the proposed CCL. This truncated process did not satisfy many in Congress, the public, or the Agency. As a result, EPA requested the assistance of the National Research Council (NRC) to design a better identification and prioritization process for the second CCL (CCL2).

The NRC convened an interdisciplinary committee that examined the strengths and weaknesses of the EPA's CCL1 process and also reviewed types of decision-making processes that could be considered for the CCL2. In a series of reports, the committee proposed a two-step process and specific methods for identifying and classifying potential drinking water hazards more efficiently (NRC, 1999a; NRC, 1999b; NRC, 2001).

This paper reviews the three major types of identification methods, lessons learned from these methods, and the NRC's rationale for its recommended process and technique. A new approach that builds on the NRC's recommendation (Parkin et al., 2002) and potential international collaborative efforts are also presented.

IDENTIFICATION AND PRIORITIZATION METHODS

When individual chemical or microbial agents are examined for risk management processes, clarification of the risk problem (usually called problem definition or initiation) is followed by a preliminary evaluative step. This first review is known as hazard identification for chemicals (NRC, 1983) and problem formulation for microbial pathogens (ILSI, 2000). The purpose of this step is to determine from existing information whether the chemical or microbial agent of concern is potentially harmful to humans and whether a more detailed risk assessment is required. This decision does not involve an intensive assessment, but often is a qualitative or semi-quantitative evaluation.

When numerous potentially hazardous substances and organisms must be evaluated and ranked for regulatory processes, risk assessment procedures for each contaminant are not feasible because of the extensive data and time required for implementation. Increasingly, regulatory programs in the United States must produce fairly rapid prioritizations of a large number of hazards, so that the greatest public health harms can be minimized and the risk reduction benefits can be achieved in a timely manner. Until the 1990s, the methods most often used were based predominantly on experts' judgments. From the 1980s, more routinized processes known as rule-based methods have been used. These methods incorporate preset scoring, weighting, and combining algorithms. However, rule-based methods have required more time than expected and have not necessarily resulted in efficient or effective prioritization processes. The NRC noted that prototype classification methods might be appropriate to consider for future prioritization tasks (NRC, 2001). This section describes and evaluates the three major approaches that may be used to identify and prioritize drinking water contaminants.

Expert Judgment Processes

The approach most often used in the United States is a set of methods known as expert judgment processes. This approach relies on techniques that combine the input of many experts using formal or informal procedures. Experts have relevant academic or technical knowledge, but typically must also have had substantial experience with drinking water issues. Participants in drinking water-related processes typically come from academic institutions, water utilities, and professional associations related to the disciplines or the public water supply industry.

The more informal methods used to combine experts' knowledge and experience involve sharing of knowledge and opinions in group meetings, technical workshops, conference calls, analyses, and/or draft documents. Examples of expert judgment approaches include the following:

- agencies' standing science advisory boards and related committees of experts (e.g., EPA's Science Advisory Board);
- time-limited and issue-specific expert committees (e.g., NRC's Committee on Drinking Water Contaminants);
- peer review panels (e.g., specially convened groups to review journal submissions or grant applications, or to oversee specific site or research projects); and
- consensus conferences (e.g., to develop better methods to track waterborne diseases).

Many times experts are convened at least once to debate their views and develop consensus. Often documents are read in advance of the debate and a group report is generated after common views have been identified. Less often, when common ground has not become readily apparent, formal methods such as the Delphi process (reviewed in NRC, 2001) have been used to finalize the group's decisions. Of the three types of identification and prioritization methods, expert judgment techniques are often the easiest to implement. They are the most familiar to experts, require few or no technical or analytical tools to support the group's process, and are relatively inexpensive. However, the outcomes of these methods have notable limitations. They may not be transparent or easy to explain to people who were not involved in the process. Additionally, the results may not be readily reproduced by a separate but similar set of experts. The decisions from this approach are highly dependent on the knowledge and views of the experts selected for the process, the group's specific scope and agenda, the dynamics that evolve within the group, and the influence of participating or contributing agency personnel. Frequently, little or no input is obtained from non-experts, leading to controversies when the experts' decisions are announced. Expert judgment processes have tended to be decide-announce-defend approaches that have often conflicted with the U.S. public's expectations.

Although still often implemented because of their simplicity and familiarity, these methods are increasingly being seen as over-used. Many times they are not the best suited for policy purposes, particularly when non-expert input needs to be integrated with expert knowledge and views or when public support and cooperation will be needed to implement the final decisions. Expert judgment processes may neither be particularly efficient nor produce the most effective and meaningful results for public policy or regulatory purposes. There is growing recognition that the disadvantages of these methods need to be more seriously considered before they are selected as the method of choice for contaminant identification and prioritization tasks.

Rule-Based Methods

These classification approaches score and then combine characteristics of contaminants in weighted algorithms to produce prioritized lists. The algorithms may be quantitative, qualitative, or semi-quantitative and result in either unique ranks for specific contaminants or broad categories for similar contaminants. Multiplicative methods are used more often than additive strategies to combine attribute values in an overall score or rank. Rule-based approaches are fairly rigid structures intended to remove subjectivity while enhancing transparency and consistency of results.

The NRC Committee on Drinking Water Contaminants reviewed 10 rule-based methods designed to prioritize hazardous contaminants (NRC, 1999a). The committee found that the results of the methods depended on the systems' purposes, the contaminants selected for ranking, the ranking criteria and weighting algorithm, and the databases chosen to determine the input information. Nearly all of the 10 methods reviewed used data on both exposure and toxicity, but none considered the impacts of the degradation byproducts of the input toxicants. All involved subjective, expert judgments, but all were less subjective than expert judgment methods. The committee determined that the rule-based processes reviewed were seen as more objective and consistent than expert judgment methods. The committee concluded that rule-based methods could provide useful preliminary screenings of a larger number of substances than can be handled using expert processes. The information needed to evaluate substances in rule-based methods may require extensive extraction of data from many sources of information. An important disadvantage of these methods is that the weights and combining methods to be used in the algorithm must be determined a priori; that is, before the method is applied to any contaminants. As a result, many regulatory processes that have relied on rule-based approaches have become quite long and controversial as experts, and sometimes stakeholders, have become engaged in intense debates about the exact weights and procedures to be used. These conflicts may occur even when the participants begin with mutually agreed-upon goals and agendas, because they begin to realize the potential consequences of the proposed weights and formulas. Similarly, the points of science policy interface in these methods may not be clear, providing more opportunities for contentious disputes to arise and stall completion of the design process. While these issues may extend the time to develop the tool considerably, the time required to implement these methods could also be lengthy. Delays and inconsistencies have often occurred, especially when the documentation for the process has lacked clear instructions about the handling of missing or imprecise data.

Many rule-based tools were developed to improve on expert judgment approaches, but have been found to have fewer advantages than expected. They are appealing because they tend to provide more rapid and useful preliminary analyses of larger sets of contaminants than do expert judgment methods. Although the implementation of rule-based ranking processes tends to be more objective and consistent than expert judgments, there may be substantial but not apparent subjectivity imbedded in them. When stakeholders discover the impacts of such hidden decisions, their trust in the regulatory ranking process may plummet, undermining the program's abilities to proceed to actions.

Prototype Classification Approaches

There are several features that set these methods apart from the other two types of approaches. Although human beings are especially adept at prototyping tasks, when they use their mental processes alone they typically cannot process a large amount of information to make decisions efficiently and consistently over time. As a result, classifications of complex entities determined through human cognition alone tend to be slow and labor-intensive. An example of prototyping can be seen in medical practice. Physicians make decisions about how to treat patients who have similar clinical diagnoses, but their treatment decisions differ from one patient to another. The doctors may not always be able to state explicitly the specific criteria they used to make those decisions, but know that experience indicates that some treatment strategies will work more effectively with some patients as opposed to others. Physicians' decision processes may be very systematic and highly effective but largely subconscious because of the large number of factors that contribute to their final treatment selections.

Prototype classification methods mimic human prototyping processes by providing a systematic, experience-based tool but have the advantage of permitting efficient analysis of a very large amount of information. This asset is often realized through the application of computerized models. These methods begin with known classifications of prototypes (obtained from historical data) that embody the kinds of outcomes one wishes to achieve. They do not rely on a classification scheme fixed a priori, but build on characteristics of past experiences by using input and outcome data from decisions already made. Just as humans do not typically discard or ignore past experience when they classify new but similar entities, these methods require acceptance of past decisions as suitable for future ones. Although prototype classification methods have been available for several decades, lack of access to high-speed computing capacity and well-accepted, standardized software limited the range of their practical applications until the 1990s.

This approach processes input variable values to construct a computerized network that discerns an algorithm, and thereby reliably maps the inputs into classification outcomes. Using data from past experiences, the network develops weights and methods to combine the variables a posteriori, removing much of the subjectivity and time involved with expert judgment and rule-based methods. The model is trained using input and output values from past experiences and is tested by using different sets of known values. Sensitivity analyses are conducted to obtain the most accurate form of the model before putting it into high-throughput use.

These methods require compilation and organization of existing information on input and output characteristics for a set of substances that is large enough to both train and test the model. Expert decisions are involved in the design but not implementation of the model, thus removing many of the negative impacts of subjectivity found in the expert judgment and rule-based methods. The design decisions involve identification of substances for which decisions similar to the ones of interest have been made, choice of data sets with information on potential input and output variables for those substances, selection of the types and nature of input and output variables to be used in the model, and choice of initial model architecture (e.g., linear, sigmoidal, or other forms of relationships between the input variables). The major activities required to obtain a stable computerized algorithm are development of the list of attributes for the model, construction of the input database, and analysis of the training set outputs to determine the most effective "cut points" between the outcome classifications.

Expert judgment and rule-based methods suffer from the problems of developing consensus for decision criteria and processes, with the latter often getting bogged down in debates about weights and formulas for combining weights. The advantages of the prototype classification approach are that it minimizes the impacts of expert judgment, does not require a priori decision-making about the weights and formulas, produces scientifically based outcomes that are reproducible, processes all input information in a non-sequential manner so that no one factor serves as a "gate keeper," allows for more than two outcome categories, and permits processing of a very large number of substances in a limited amount of time. Additionally, by using only fundamental characteristics of substances as input variables, proactive identification of hazards well beyond substances related to already known or suspected risks becomes possible. The primary advantage of prototype classification methods that distinguishes them from rule-based methods is the a posteriori determination of weights on the basis of past experience.

The disadvantages of prototyping methods are that they may require a lot of data for many substances; involve some assumptions and judgments to design the model; may result in an over-designed tool (i.e., a model that is more complex than necessary to make the desired decision); and may not be readily understood by people who have not been involved in the design decisions. However, expert judgment and rule-based methods also involve many aspects that are not transparent, not documented, and may not even be recognized by the persons who construct the processes and outputs. Specifically, the determination of weights and consequences of design decisions may only be superficially transparent.

When using prototype classification schemes, the goal should be to develop the simplest, most robust model that produces the type of decision required; e.g., the identification of drinking water contaminants that should be considered further for regulation. The accomplishment of this goal through simple models not only enhances the ability to explain the input-output relationships, but also reduces the amount and complexity of data and documentation needed to replicate the results.

The major forms of prototype classification schemes are machine learning classification systems, clustering algorithms, neural network models, and hybrids of these three. Only neural network modeling will be discussed further here, because this is the set of models that have recently been pilot tested (NRC, 2001) and are being developed (Parkin et al., 2002) for the proactive identification of drinking water contaminants. Although this method has been used widely for other purposes (e.g., clinical and financial decisions), it has not been previously used for environmental hazard prioritization processes.

NEURAL NETWORK MODELS

This method typically involves several to many input variables processed in a forward configuration to generate a fairly small number of outcome classifications. To use neural network models the designer chooses the model's architecture; that is, determines the input variables and scoring systems to be considered, the number of neurons (or decision points), the organization of the model, the transfer functions that operate at each node, and the methods to analyze the sensitivity and set the "cut points" between the outcome categories. The method provides the capacity to capture linear and non-linear dependencies and constructs multiple, complex relationships that are not readily apparent in human prototyping activities (Garson, 1998). The method can produce simple, single-node models or highly complex models, such as are used to predict credit-worthiness in financial settings. However, the simplest model that effectively produces reliable outcomes is preferable.

The NRC committee recommended neural network models to the EPA for more rapid identification of the set of substances that should be considered for regulatory purposes. This recommendation was based on several important observations and decisions. The committee recognized that EPA's past reliance on expert judgment and rule-based methods has not produced results in an efficient, open, or reliable manner. The time and staff resources available to EPA to prepare the CCL2 were important factors that led the committee to design the simplest process that would achieve a credible set of outcomes. The committee agreed that the process had to be science-based, systematic, defensible, and transparent. Finally, the group recognized that past prioritization processes used by EPA had evaluated chemicals and microbial pathogens separately, but felt that these two sets of contaminants needed to be integrated in one streamlined process. Thus, the committee decided to design a tool that would have the ability to evaluate both chemicals and microbial pathogens with the same input variables. As a result, programmatic priorities could be set across, and not just within, broad or superficial contaminant categories (NRC, 2001).

The committee developed and recommended a two-stage identification process: going from a "universe" to a shorter initial list before screening again to obtain the final list of substances for regulatory consideration. From all existing chemicals, microorganisms, radiological and other substances (hundreds of thousands of substances), there is a smaller subset (tens of thousands of substances) that constitutes the universe of substances that may be found in drinking water. This universe was the committee's starting point for the recommended two-step process (Figure 1).

First, the universe would be screened to create a Preliminary Contaminant Candidate List (PCCL) containing no more than a few thousand substances. The winnowing required for this step may be accomplished through the use of simple screening criteria and limited expert judgment. For example, the data needed to implement this screening step for chemicals would include information on their use, location, and physical characteristics. Substances on the PCCL would be those that are known to occur or potentially occur in drinking water and those that are known or suspected to cause adverse health effects. The combination of

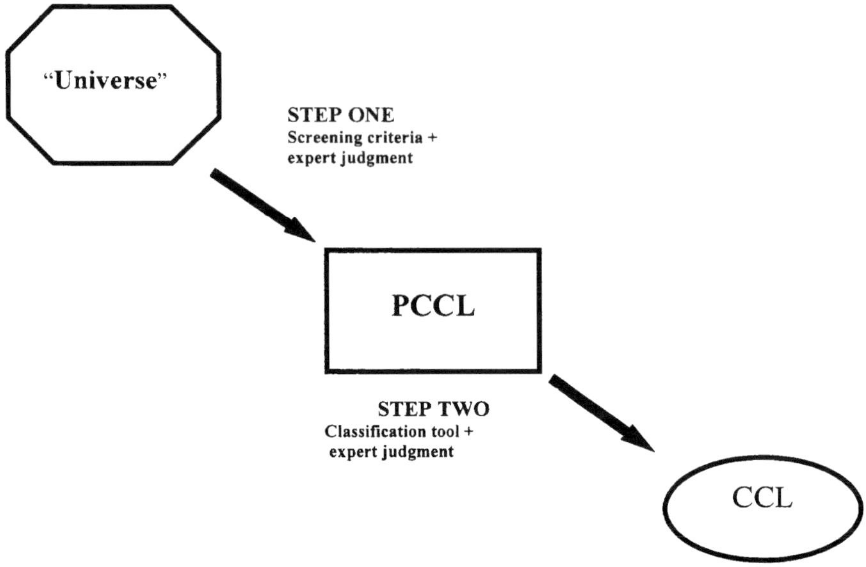

FIGURE 1 Recommended two-step process for identification of contaminants for regulatory consideration (As adapted from NRC, 2001). PCCL is the Preliminary Contaminant Candidate List, and CCL is the Contaminant Candidate List.

these two criteria would result in four categories of substances that would be placed on the PCCL but treated equivalently from that point (Figure 2).

In the second step, the PCCL would be assessed to determine which substances (potentially about one hundred) should be placed on the CCL and therefore be considered further for regulatory purposes. The committee recommended the use of neural network models for narrowing the PCCL to the CCL (NRC, 2001). The committee found the development of the model fairly simple and time-efficient. It required committee consensus on the few input variables that would be used, the scoring schemes for each variable, and the datasets that would be used to score the substances' attributes. A few committee members individually conducted the additional model development and implementation activities.

Consensus on the variables to be used as inputs, describing the potential occurrence and health effects of substances, was obtained very rapidly. There were five input variables: three for occurrence (prevalence, magnitude, and persistence-mobility) and two for health effects (severity and potency). Upon initial review of the original six attributes selected by the committee, persistence and mobility were found to be so highly correlated that only one score for these

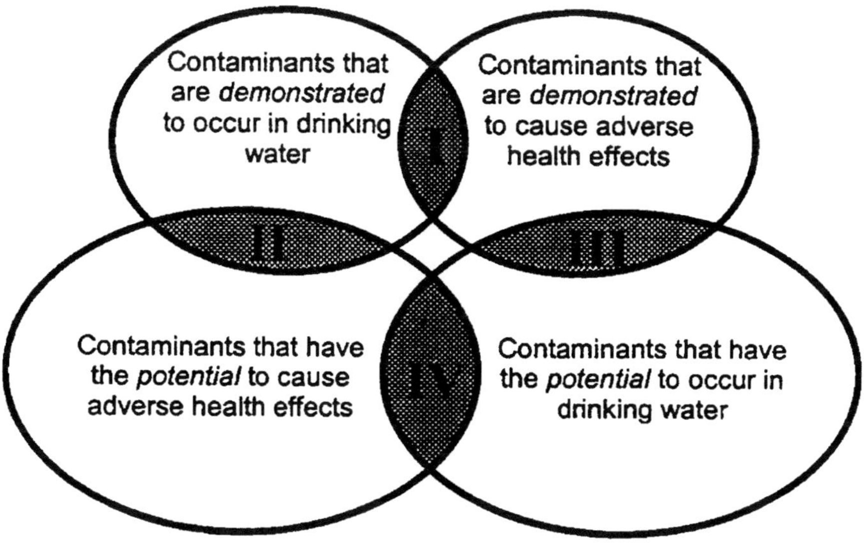

FIGURE 2 Conceptual approach to identifying contaminants for inclusion in a Preliminary Contaminant Candidate List (PCCL) (NRC, 1999a).

two attributes was necessary in the model. Relevant information about chemicals was known to exist in well-recognized, peer reviewed databases and could be readily accessed for the scoring task (NRC, 2001).

The committee then broke into subcommittees to develop scoring frameworks for microorganisms, occurrence, and health effects. Microorganisms were evaluated separately, because existing databases did not include them, and fundamental decisions were needed to determine whether pathogens could be included in our modeling exercise. Ultimately, a separate chapter was written about microorganisms proposing a new approach to obtain the data necessary to score them simultaneously with chemicals. The final frameworks for scoring occurrence and health effects were based on a number of scientific issues. The scores for the five attributes ranged from zero or one through ten (NRC, 2001).

A list of 85 contaminants (chemicals and microorganisms) that do or could occur in drinking water was prepared based on past regulatory decisions. Using the subcommittees' draft scoring systems and existing databases (developed by EPA, the U.S. Food and Drug Administration, and the National Library of Medicine) with the necessary information for the substances, the input database was created in about two weeks by a toxicologist, with limited input from a few other committee members who had more detailed knowledge of complex chemicals or the design of the databases. Only one draft scoring scheme (that for severity) had to be revised after the first few attempts to use it.[1] Another committee member selected the modeling software (Matlab and Matlab Neural Network Toolbox of Mathworks, Inc., in Natick, Massachusetts), chose the training algorithm (conjugate gradient method), and used the substances' input scores to train and test the neural network model. A sensitivity analysis of the initial training set results showed that there had been limited misclassification of past regulatory decisions. The cut point for the outcome variable ("consider further for regulation" or "do not consider further") was revised based on the committee's review and discussion of these findings. The revised cut point produced highly accurate and reliable results in the model rerun. Data for 63 regulated and 17 unregulated substances were used to train the model, and information on five substances was used to test it.

The final model had two layers connected in a feed-forward configuration (Figure 3). The first layer was hidden and included two nodes, each of which involved a sigmoidal (simple, non-linear) transfer function. The second layer was a linear function that produced the two outcome categories (consider further or not). Although more complex models could have been constructed, a comparison of this model's outputs with those of a strictly linear model revealed that the neural network model yielded superior results with a minimum of additional complexity. The two-layer model met the goal of achieving effective outputs with the least complex model, so the committee recommended this simple neural network model form for the PCCL-CCL screening step.

Time and resources, however, severely limited the committee's ability to develop the model to the extent needed for regulatory purposes. Recognizing this, the NRC advised EPA to review the committee's fundamental decisions, determine what would be needed to support a public process to implement the neural network model approach, and test the model with several datasets before using it for obtaining the CCL2 (NRC, 2001).

Following EPA's acceptance and review of the NRC committee's report, several committee members oriented EPA staff to the details of the neural network model development process. From winter 2002, staff members began making the fundamental decisions necessary to design EPA's version of a PCCL-CCL neural network model. They also began compiling the necessary input databases so that the model could be developed and CCL2 published by the mandated deadline of February 2003. However, by fall 2002 the agency determined that additional time was needed to implement means not only to develop a model but also to establish an effective stakeholder process. The NRC committee pointed out the importance of this input and urged the EPA to include as much public comment

[1]The range of health outcomes scored was found to be too narrow. All substances were receiving scores at the upper end (8-10), which did not provide enough spread in values for modeling purposes. When the outcomes were redefined (see NRC, 2001), the model performed very well.

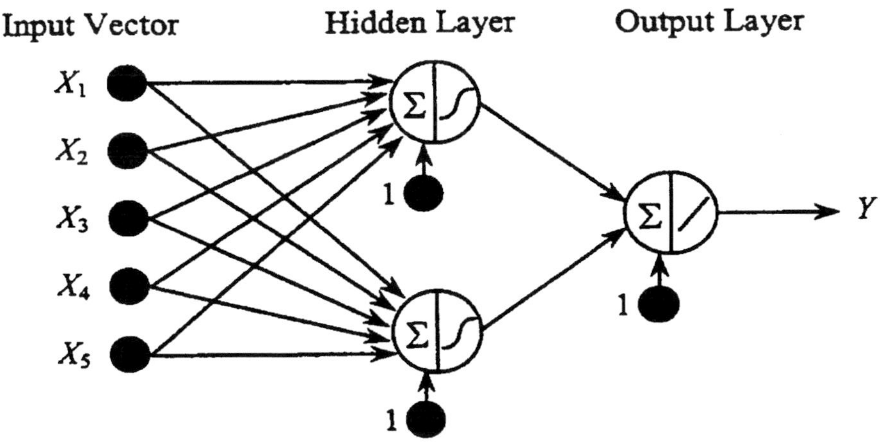

FIGURE 3 Multi-layer neural network model used for contaminant classification (NRC, 2001).

as possible. However, with the time available, EPA was not able to fully meet the participatory ideals described by the committee (NRC, 2001). EPA is now targeting the 2007 regulatory timeframe for the modeling and stakeholder processes.

LESSONS LEARNED

Although they have been extensively used, expert judgment and rule-based processes do not have the capacity to screen large numbers of potentially hazardous substances to protect public health in a timely, efficient, and consistent manner. Concerns about the possibly high impacts and low transparency of subjective decisions in these two approaches have reduced their value for regulatory processes. As a result, prototype classification schemes, particularly neural network modeling, are now being developed and tested to determine their advantages and disadvantages for identifying and prioritizing contaminants. Already important lessons have been learned about this approach for drinking water contaminants.

Prototyping may reduce the time and number of controversies involved in developing consensus on the key attributes of substances and the sources of data to score those attributes. Scoring schemes may require some consensus building, but are unlikely to entail extensive decision processes. For example, although the NRC committee's draft severity scale was too limited to distinguish substances' health impacts effectively, the scaling problem became apparent in scoring only a few substances and was readily improved and reapplied successfully. Fairly general attribute data appear to be sufficient to sort substances into those that merit further regulatory consideration and those that do not. An important part of developing a neural network model is examining the potential correlations and culling out highly correlated attributes so that the final set of attributes used in the model only includes those that are necessary to produce reliable outcome values.

Data sufficient to score attributes is readily available and does not require highly technical or time-consuming evaluations to determine appropriate scores. The choice of databases to be used in the scoring step should relate to the quality and relevance of the information to the attributes to be included in the model. For example, in a model focused on protecting public health, data such as ambient water information has less immediate relevance than tap water data for scoring occurrence. One important limitation is the lack of appropriate attribute data in a centralized database for microorganisms. The necessary data may become available in the next few years, but were not ready to any significant extent for the CCL2.

A sufficient number of substances need to be scored and used to ensure a robust neural network model. Five times the number of attributes to be used is the minimum number of substances that should be used to train the model. A small number of substances can be processed to test the model, but several data sets should be used to conduct repeated tests and thereby ensure the model's stability before broad application.

The importance of proactively identifying substances for regulatory purposes cannot be understated. Not only will more efficient and comprehensive screening processes produce greater public health protection, but they will also provide more time for new risk management strategies and technologies to be developed.

A NEW DIRECTION

Drinking water utilities not only attempt to anticipate regulatory requirements, but also must address public concerns about actual or potential drinking water contaminants. In an effort to develop a tool so that emerging contaminant concerns can be foreseen more readily, the American Water Works Association Research Foundation (AWWARF) funded a project to identify contaminants proactively (Parkin et al., 2002).

This project is focused on combining scientific data (as in the NRC model) and public concerns in a neural network model that will provide early indication of substances that may require more detailed attention and review. It is based on the premise that substances may emerge as public concerns from the public's perceptions of substances rather than from scientific knowledge alone. The conceptual foundation for this project is several decades of scientific investigations into the formation of expert and non-expert risk perceptions (Fischhoff et al., 1997). It is well known that the factors that influence the perceptions of these two groups differ in important ways (e.g., Morgan et al., 2002).

The investigators are determining the input attributes (and appropriate data sources) that will result in four distinct outcome categories: 1) substances that are of both scientific and public concern, 2) substances that are of neither scientific nor public concern, 3) substances that are of scientific but not public concern, and 4) substances that are of public but not scientific concern (Figure 4). Although the work on this project is in initial stages, evaluation of the sources and types of information available to evaluate public concerns has revealed some themes and directions that will shape the final model. The results of the project will be reported in late 2003 and are expected to be of interest to utilities and regulatory agencies.

PROPOSED COLLABORATIONS

At this early stage of neural network model development for drinking water contaminant identification, the opportunities to test the method in a range of regulatory and practice settings have not been adequately considered. As more organizations attempt to design and use neural network (or other prototype classification) models to protect public health, it will become possible to evaluate the impacts of differing designs, decision-making processes, and societal

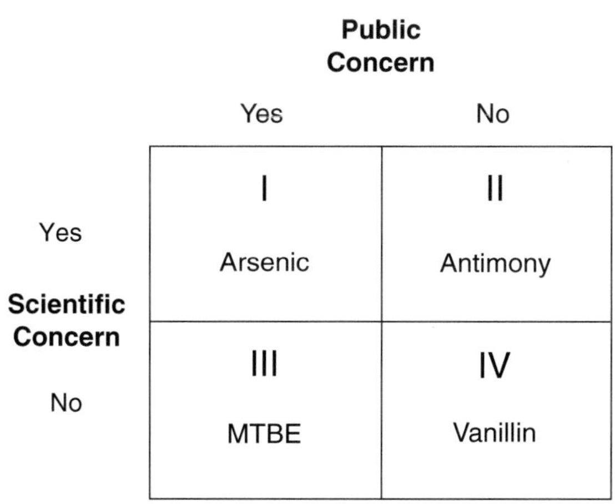

FIGURE 4 Proposed outcomes for a neural network model with examples of substances in each category (Parkin et al., 2002).

I. Scientific and public concern
II. Scientific concern, but no public concern
III Public concern, but no scientific concern
IV. No scientific or public concern

contexts. The lessons learned from these efforts need to be reported in the peer-reviewed literature so comparisons can be made. Ideally, studies should be designed so that the impacts of differing model development processes and decisions can be evaluated systematically, and scientifically derived lessons can be obtained.

Collaboration among experts with differing but complementary backgrounds has taken place in the NRC's committee and is now occurring at the EPA and in the AWWARF project based at The George Washington University in Washington, DC (Parkin et al., 2002). Because of the range of knowledge, data sources, and techniques that these teams bring to the model development activities, their efforts are likely to result in a broader range of insights and more robust neural network models. Similarly, collaboration of experts from different regulatory settings would provide new opportunities to understand when and how neural network models may be most useful as decision support tools for regulatory purposes.

A range of prototyping techniques should be considered, developed, and tested in a variety of contexts. Formal collaborations among experts from different settings would advance capabilities and insights about the tools' value for contaminant identification tasks designed to meet a variety of goals. Diverse settings would offer an opportunity to design a scientifically based social research initiative in which the impacts of differing decision-making processes and modeling choices on the techniques and the regulatory goals can be examined and reported in a more rigorous and credible manner.

CONCLUSIONS

In conclusion, expert and rule-based methods have been found to be too limited for efficient, proactive drinking water contaminant identification and prioritization processes for regulatory programs. During the 1990s, new computing and software capabilities became available and have produced stable prototyping classification methods that can be more widely used to address drinking water issues. These methods may improve both the speed and effectiveness of contaminant assessment, but many lessons need to be learned about the tool's utility in a variety of settings. Experts' knowledge about the strengths and weaknesses of the approach will grow more quickly and advances will be made as experience is gained in diverse settings. With deeper understanding of prototyping methods and their practical value, regulatory program managers will be able to make better decisions about how to identify and rank hazardous substances. The large number of new chemicals being made and microorganisms evolving makes proactive public health protection more urgent, and innovative decision-making tools and methods more critical. Prototyping classification techniques offer regulators a new, efficient and reproducible way to achieve this goal, possibly with fewer controversies. The lessons learned in the next decade about this set of techniques are expected to open up opportunities for more rapid and appropriately focused public health initiatives.

ACKNOWLEDGEMENTS

The author acknowledges the contributions made by her colleagues on the NRC Committee on Drinking Water Contaminants that serve as the partial basis for one of her current research projects, the American Water Works Association Research Foundation Project #2776. She also gratefully recognizes the technical assistance of Lisa Ragain and Martha Embrey, research staff of the Center for Risk Science and Public Health, School of Public Health and Health Services, The George Washington University, Washington, DC.

SELECTED BIBLIOGRAPHY

Balbus, J.M. and M.A. Embrey. 2002. Risk factors for waterborne enteric infections. Current Opinion in Gastroenterology 18(l): 46-50.

Bourbigot, Marie-Marguerite. 2000. Access to healthy water and an efficient wastewater treatment system. World Conference on Science. Science for the Twenty-First Century: A New Commitment. Paris: United Nations Educational, Scientific, and Cultural Organization. Available at http://unesdoc.unesco.org/images/0012/001207/120706e.pdf.

Canadian Standards Association (CSA). 1997. Q850 Risk Management: Guideline for Decision-Makers. Toronto.

Carson, R. 1962. Silent Spring. Greenwich, Connecticut: Fawcett.

Fischhoff, B., A. Bostrom, and M.J. Quadrel. 1997. Risk perception and communication. Pp. 987-1002 in Detels, R., W. Holland, J. McEwen, and G. Omerm, eds. Oxford Textbook of Public Health. London: Oxford University Press.

Garson, G.D. 1998. Neural Networks: An Introductory Guide for Social Scientists. Thousand Oaks, CA: SAGE Publications.

International Life Sciences Institute (ILSI). 2000. Revised Framework for Microbial Risk Assessment. Washington, DC: ILSI, Risk Science Institute.

Morgan, G., B. Fischhoff, A. Bostrom, and C.J. Atman. 2002. Risk Communication: A Mental Models Approach. Cambridge: Cambridge University Press.

National Research Council (NRC). 1983. Risk Assessment in the Federal Government: Managing the Process. Washington, DC: National Academy Press.

National Research Council (NRC). 1997. Safe Water from Every Tap. Washington, DC: National Academy Press.

National Research Council (NRC). 1999a. Setting Priorities for Drinking Water Contaminants. Washington, DC: National Academy Press.

National Research Council (NRC). 1999b. Identifying Future Drinking Water Contaminants. Washington, DC: National Academy Press.

National Research Council (NRC). 2001. Committee on Drinking Water Contaminants, Classifying Drinking Water Contaminants for Regulatory Consideration. Washington, DC: National Academy Press.

Parkin, R. 2002. Strategies for identifying emerging drinking water contaminants and communicating to address public concerns. American Water Works Association Research Foundation (AWWARF) Research Project 2776. Denver, CO: AWWARF. Available at http://www.awwarf.com/research/.

Presidential/Congressional Commission on Risk Assessment and Management. 1997. Risk Assessment and Risk Management in Regulatory Decision-Making: Final Report, Volume 2. Washington, DC.

Safe Drinking Water Act (SDWA) Amendments of 1996, 42 U.S.C. 300 (1974).

United States Environmental Protection Agency (EPA). 1987. Risk Assessment Guidelines of 1986. EPA-600/8-87/045. Washington, DC.

United States Code Annotated. Reorganization Plan No. 3 of 1970. 91st Congress, Second Session, Vol. 3.

World Health Organization (WHO). 2002. 2000-2001 Work Plan. Available at http://www.who.int/docstore/water_sanitation_health/General/workplan.htm.

Integrated Approach to Water and Wastewater Management for Tehran, Iran

Massoud Tajrishy and Ahmad Abrishamchi

ABSTRACT

Tehran, with a population of over 7 million, is experiencing perhaps the fastest urban development of all Asian cities. The population growth in the next decade will place immense demands on the city's water resources. Tehran is one of the largest cities in the world that lacks an adequate wastewater collection and treatment system. Most of the city's wastewater is disposed under the ground, without any treatment, through the use of injection wells to recharge the groundwater. This type of disposal is unique and has caused some water supplies to be polluted, raised the water table, and degraded surface water channels. In this article, analysis is made of the increasing potable water demand, resulting from continuous population growth; inadequacies in the water resources supply that are causing a chronic potable water shortage; and pollution of raw water due to inappropriate disposal of wastewater without adequate treatment. The extent of these increasing demands when placed against the constraints of water, time, funds, and the lack of an integrated approach to manage drought impact and pollution of groundwater, confirms that the city faces a serious water and wastewater crisis.

INTRODUCTION

The tremendous speed of population growth (often doubling within only 10-20 years) in many major cities of developing countries is much faster than the speed at which city authorities can increase services. Growing cities often destroy their own water sources, while the new sources farther away rapidly become insurmountably costly to use. Such is the case in Tehran. As is happening in many megacities worldwide, metropolitan Tehran is experiencing rapid urban growth, with serious concerns being raised regarding the environmental sustainability of this development, and the potential detrimental impacts on the quality and quantity of groundwater resources.

The city of Tehran, with an existing population of over 7 million people, covers an area of approximately 730 km^2 and lies within the Tehran basin on the semiarid plains to the south of the Alborz Mountains, with a varied terrain; from steep hilly areas in the north to plains in the south. The average slope of the area ranges from 1.3 to 5 percent. The mean annual precipitation is only 250 mm and occurs mainly during the winter and spring. No rivers of any size pass through the basin, but groundwater is contained in the extensive alluvial aquifer that underlies the basin.

What is needed is to move away from the technical-fix dominated, and largely supply-oriented management structure of water resource management. The focus has to be extended from blue water flow to incorporate also green water issues, and from water quantity to incorporate water quality as well.

An integrated approach is necessary for environmental management and water management of megacities like Tehran. Urban planning, as it relates to water, should encompass the integration of the physical land beyond the city limits, considering both the river basin where the city is located and the surrounding region affected by and interacting with the city. Planning should also incorporate a multisectorial framework. All sorts of interdependency linkages and implementation barriers need to be addressed in an overarching and integrated manner. The conventional setup of sectoral water management institutions is not able to cope with the present water problem facing the city of Tehran. The solutions to these problems require an integrated approach to water, land use, and ecosystems, addressing the role of water within the context of social and economic development and environmental sustainability.

Basically what is spoken of as a water problem in the city is not solely a water problem but a societal problem; the main task is to master the driving forces, to build up the balancing forces, and to develop competent governance systems. Megacities like Tehran are not able to cope with the growth of the suburbs, and in many cases development has gotten out of control. The unbalanced population growth and reliance on long distance water transfer are among the challenges that the city faces.

Problems that are facing water resources management in Tehran can be summarized as increase in demand and waste production due to population growth and socioeconomic development; decrease in availability of water per capita; high losses of urban water; and local depletion and pollution of surface and groundwater. Urban water management in this city will fail without a holistic and integrated view.

SOURCES OF WATER

The traditional method of providing water in Tehran, as well as in other parts of Iran and the Middle East, is by means of "qanats," which are small-diameter, hand-excavated tunnels that slope gently upward from the plain to tap the groundwater at higher levels near the foothills of the mountains. Before 1927, 26 strings of qanat with a total flow rate of 700 liters/second had been supplying the city's water needs.

The first supply to the city from a surface water source was commissioned in 1933, bringing water by aqueduct from the Karaj River some 40 km to the west. Since that time, the population of Tehran has grown from some 300,000 to more than seven million people.

Water is currently one of the major constraints on the city's development. The city gets its water from both surface water and groundwater. Surface water supplies have been augmented by the construction of the Amirkabir Dam on the Karaj River, duplication of the original aqueduct, and the construction of the Latiyan dam and tunnel diverting water from the Jajrood River east of the city. In 1978, the construction of the River Lar project, which consists of a major dam, was commissioned for several purposes: regulating the flows of the Lar River, providing a storage controlled flow, releasing water down the river to Mazandaran, and creating a diversion tunnel system transferring Lar water to the Latiyan Reservoir for onward conveyance to Tehran. Hydroelectric power stations are included in the diversion system to utilize the appreciable head between the two reservoirs.

The annual exploitation of subterranean water for supply is increasing and currently is on the order of approximately 400 Mm^3, which is drawn from 200 deep wells with an average depth of 130 m. Approximately 40 percent of the population of the city relies on groundwater for its drinking water needs. The quality of drinking water available from Tehran's aquifers is generally very good, although the aquifers are susceptible to contamination. Additionally, since the 1998 drought, 20-30 Mm^3 of water is pumped annually from wells adjacent to the Jajrood River into the tunnel reaching Tehran's No. 5 Water Treatment Plant. The rest of the city's municipal demand is met by water from three reservoirs located in nearby watersheds (Figure 1). The distribution of the water supply for the city of Tehran during the last 25 years is shown in Table 1 and Figure 2.

WATER USE

From 1955 to 1995, over about 40 years, the water consumption in Tehran rose from 10 Mm^3 to more than 800 Mm^3, an increase of eighty times. Tehran's water demand began to expand rapidly as the Iranian economy grew. This expansion became particularly rapid during the 1970s and 1980s due to the nature of the rapid economic growth, causing industry and population to be

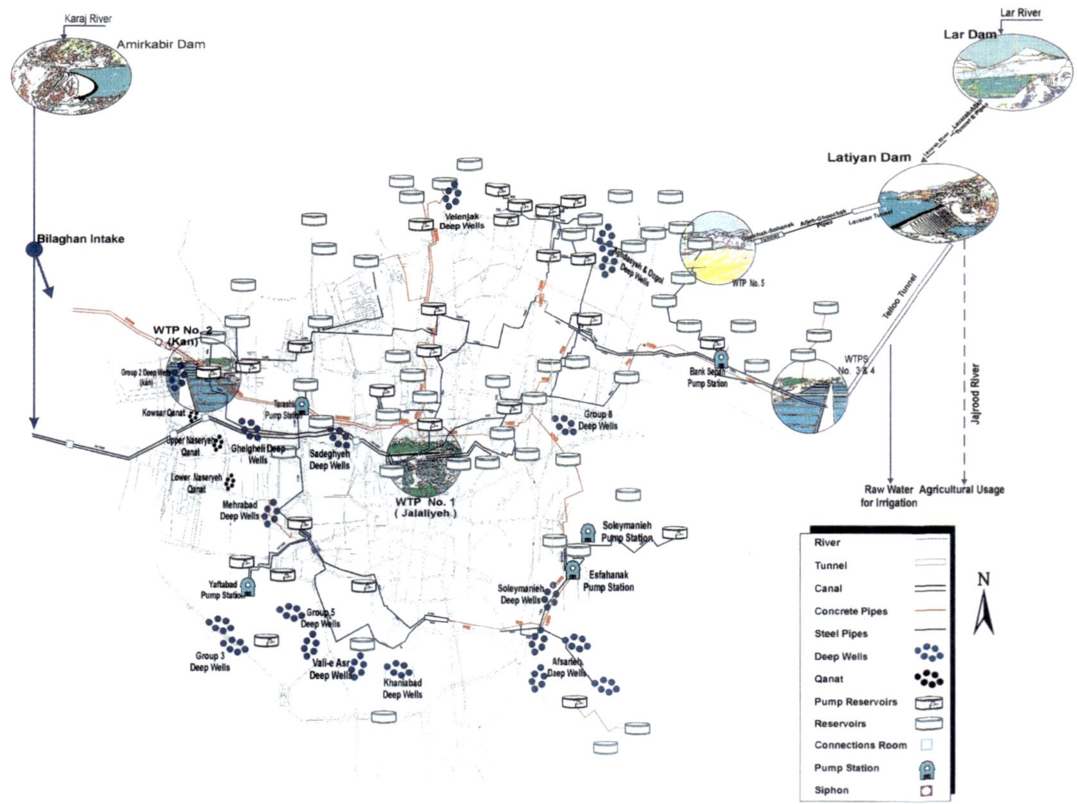

FIGURE 1 Schematic map of Tehran drinking water reservoirs, deep wells, water treatment plants (WTPs), and dams.

TABLE 1 Proportion of Water Supplied by Groundwater (GW) and Surface Water (SW) for the City of Tehran

	1975		1985		1995		2000	
Source	%	Mm^3	%	Mm^3	%	Mm^3	%	Mm^3
Karaj River (Amirkabir Dam)	60	212	57	310	43	320	29	270
Latiyan and Lar Dam	30	108	33	180	36	290	29	270
Groundwater	10	35	10	55	21	170	42	390
Total	100	355	100	545	100	810	100	930

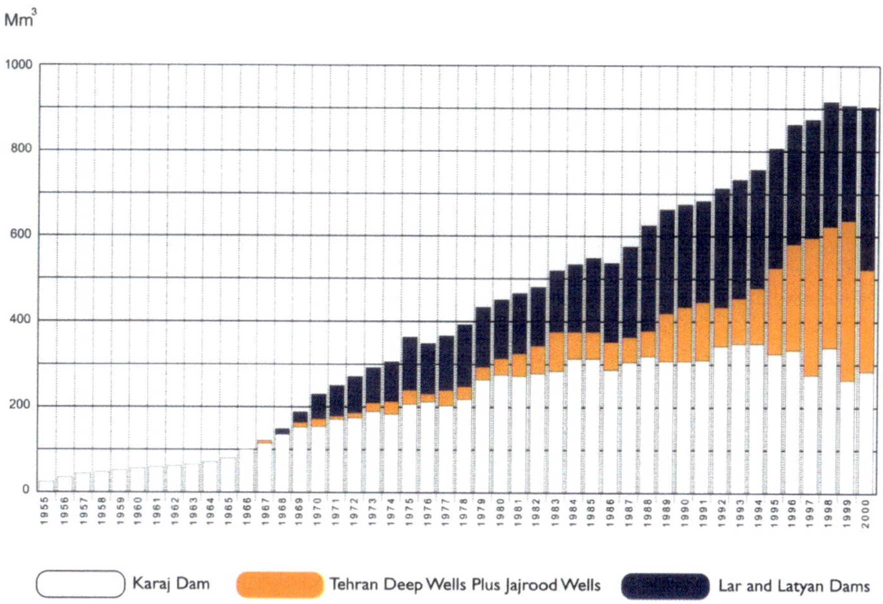

FIGURE 2 Tehran's water supply resources during 1955-2000.

concentrated in the capital, and the massive migration of population after the Islamic revolution in 1978.

Population and water usage during the last 35 years is given in Table 2. The population of Tehran has grown steadily due to the intense centralization of the majority of the political and economic functions of the country, as well as spreading unemployment in other provinces. Recent studies show that the population

TABLE 2 Trends in Population Growth, Total Water Usage, and Daily Per Capita Water Usage of the City of Tehran

	1965	1975	1985	1995	2000
Population (1000)	2700	4500	6000	6600	7200
Total Water Usage (Mm^3)	96	355	545	810	930
Liters/Person/Day	97	160	250	335	350

migration and birth rate add to the city's population by 300,000 and 70,000 persons per year, respectively. As illustrated in Table 2, although population has increased by about 2 percent yearly, water usage (liters per capita) has increased at a rate of about 5 to 6 percent annually.

During recent years water consumption has risen above 350 liters/person/day. If the population continues to rise at the same rate (about 2 percent annually) as it did from 1995 to 2000, Tehran's population alone will reach about 10 million by 2015. If the high population growth of the last 25 years slows down dramatically and population migration stops, the rate of population growth may slow to 1.7 percent, which will result in a population of above 10 million by the year 2020.

In the years 2010 and 2020, the volume of water consumption in Tehran is projected to reach 1,100 and 1,400 Mm^3, respectively. Based on current water usage and anticipated population growth, the water shortage is expected to grow to about 100 Mm^3 in the years ahead; about 400 Mm^3 by year 2015, and about 600 Mm^3 by year 2020. Most of the water is expected to be supplied by groundwater pumping (which will be discussed later and which has its own problems) and the transmission of water from further distances.

WASTEWATER

The city of Tehran is located on an alluvial plain. The alluvium is composed of sand, ballast, and clay, with high permeability in the northern areas of Tehran, due to the concentration of sand and ballast, and poor permeability in the south, due to clay content. The water table north and south of the city is at a depth of approximately 70 m and 3 to 4 m, respectively. Because of the huge costs involved in a rapidly expanding city, the development of the sewer system did not keep pace with the construction of buildings and highways.

The city lacks municipal sewage facilities; hence, the only method of sewage disposal for domestic waste is through seepage pits and leaching cesspools. There is a potential for waste from these leaching cesspools to leach into the underlying aquifer. The extensive use of cesspools in Tehran has caused the water table to rise (Figure 3). The average distance between the water table and the bottom of

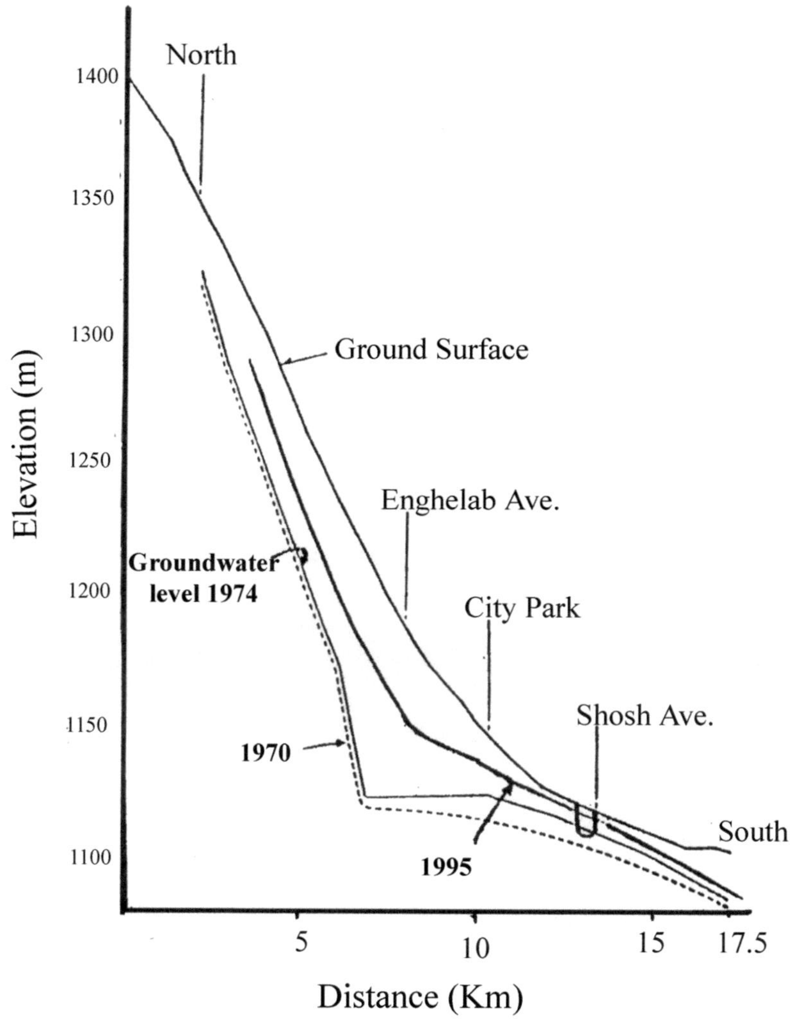

FIGURE 3 Historical groundwater level changes in the city of Tehran.

the seepage pits within the city limit is about 20 m. The level, however, varies from the north to the south.

Southern seepage pits and leaching cesspools are more likely to break through to the aquifer because of the shallow water table and because of saturated soil conditions during periods of heavy rainfall. In the south, where soil content is high in clay, the water table is close to the surface. In these areas, a network of pumping wells and a proper collection system have been built to transport and dispose of the pumped groundwater to keep adequate distance between the seepage pits and the water table. In some areas, these facilities have resulted in differential soil settlement, causing additional structural and environmental damage.

Sixty to seventy percent of Tehran's water ultimately penetrates the ground because of the lack of a sewage system. Consequently, the water table is rising rapidly throughout the city; the average rate is approximately 1 to 2 meters per year, but in some areas the water levels have risen 10 meters in only four years.

Contamination from rising groundwater and wastewater intrusion are significant factors potentially limiting groundwater use. The direction of groundwater flow is basically toward the south and southeast, with a mean velocity of approximately 0.2 m/day.

A sewage collection and treatment system has been planned and designed in two stages. The first phase, which is expected to be completed by the year 2015, will cover 15,000 hectares of the city, consisting of 10,000 hectares in the south and about 5,000 in the north, and serving about 2 million people. The second phase of the project will cover the remaining part of the city and has an unknown completion date. The project calls for two treatment plants to be located in the south and west sections of the city. Effluent from these plants will be used for irrigation while the sludge from the treatment works will be used as fertilizer.

PROPOSED SOLUTIONS TO WATER SHORTAGE

Curtailment of population growth, water reuse, administrative and technical measures to obtain savings in water consumption, a reduction in losses in the distribution system, and better management of local water resources are proposed as solutions to Tehran's water problem. These measures are preferred over importing water from outer basins, which results in unnecessary expenditure and ultimately transferring the water problems to other areas.

Population Control

The explosive population growth trend for the Tehran Metropolitan Area (TMA) may be curtailed by means of short-term and long-term policies that increase the quality of living and working conditions in other parts of the country and at the same time reduce the economic attractiveness of the capital. At present, 20 percent of the country's population lives in TMA. The current scheme of hierarchical centralization should be abandoned in favor of a more balanced outline that will foster increasing development beyond the basin of Tehran. Figure 4 shows the formation of new towns (or satellite towns) from 1950 to 1990, which helped to control the growth of Tehran, as the influx of immigrants was concentrated in the outside areas.

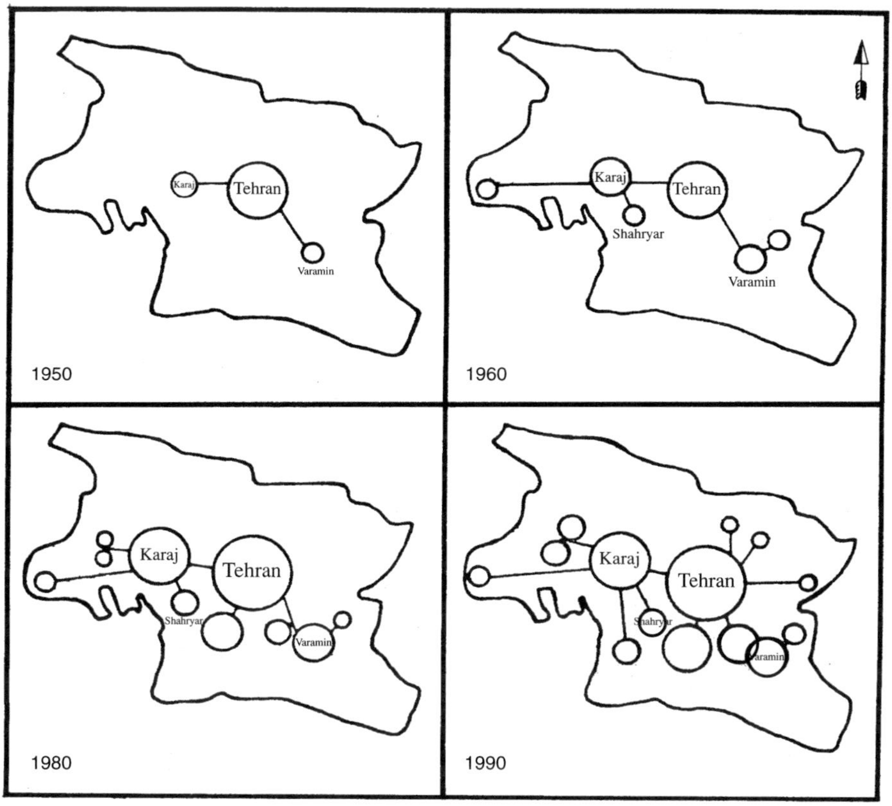

FIGURE 4 Population increase in the suburbs of Tehran from 1950 to 1990.

Wastewater Reuse

Reclaimed wastewater is a reliable source of water that must be taken into account in formulating a sustainable water policy. Recycling and reuse of wastewater presents a very important option for water planners in Tehran. Reuse is a promising avenue for solving water resource problems by providing additional quantities of water. Reuse is, therefore, a water management option that should be carefully considered in the future planning of Tehran's water, and utilized to its full potential. City wastewater can be reused both in the city and in the peri-urban area as a potential resource for food production, both for farmers themselves and for the urban market. The health hazard risk involved can be mitigated by simple and available treatment methods. Integrating the wastewater reuse component into the city's strategic water plan will give the potential to fully utilize more than 600 Mm^3 of wastewater produced yearly. This makes water reuse within the city a promising alternative. If there are any barriers to adopting water reuse, the barriers must be reshaped to cope with an era of better utilization of limited resources, an era of living with what is physically available.

Landscaping and Recreational Use

In semiarid areas like Tehran, water is in short supply for irrigating natural vegetation, landscaping, and park areas. The city of Tehran has about 20 km^2 of parks that use freshwater for their irrigation. Many of these parks are located close to satellite wastewater treatment plants from which the outflow is currently being discharged to seepage pits and surface stormwater channels. In one study, it has been shown that about five parks located in Tehran (total 5,000 ha) use 30 Mm^3 of freshwater that can be replaced by reclaimed wastewater from treatment plants close to them. Outflow from treatment plants can even be used for landscape impoundment and groundwater recharge in the eastern part of the city, which is under development.

Industrial Reuse

Water efficiency is very low in the industrial sector of the country, and there is still not enough emphasis on water recycling and reuse. Municipal wastewater after treatment can be reused for cooling and processing water in industry. This has become an established practice in many countries. The greatest potential for industrial water reuse in Tehran is to supplement or replace the potable water demand of Ray Petrochemical Complex located south of the city. This complex uses more than 10 Mm^3 per year of potable water mainly for industrial processes. Treated municipal wastewater effluent from the south wastewater treatment plant can be used for a significant fraction of the water requirements of Ray Petrochemical Complex. Other industries in the western part of the city (Karaj Industrial Park) can use reclaimed wastewater for direct evaporative cooling, indirect refrigeration (food processing), or for in-plant transport and washing.

Irrigational Reuse

Historically, the reuse of irrigation water has been quite common in the Varamin-Garmsar basin (south of Tehran) after reduction of Jaj-e-Rood River outflow. The southwestern part of the basin, after construction of Amirkabir dam, experienced a lowering of the water table due to a decrease in aquifer recharge. Irrigation by wastewater has a great potential that has not yet been considered. One particular advantage to irrigation reuse is that almost no advanced wastewater treatment is necessary. The biological material remaining in the wastewater effluent can provide valuable nutrients for crops.

Wastewater injection into local aquifers at selected sites within the basin should further be investigated in order to improve aquifer recharge, which has decreased continuously because of growth of urbanized areas and drainage of increasing volumes to the Tehran basin.

Administrative and Technical Measures

Conservation is achieved through a combination of efforts that includes legislation, pricing, incentives, coalition building, research, and education. While figures of water consumption per capita in the industrial countries are almost stable as a consequence of water-saving techniques in industry and households, the water consumption in the city of Tehran is still increasing.

Conservation efforts must be focused primarily on information and education programs. A strategy should be adopted to reduce per capita water use by 20-25 percent within 5 years. It is estimated that if conservation efforts could decrease water use by 20-25 percent by the year 2010, the current water supply could be extended by 10 years.

Establishment of a water conservation planning board is needed, and this board needs to promote implementation of best management practices (BMPs) over the next 5 years. The BMPs are incentive pricing and billing, water measurement and accounting systems, information and education programs, distribution system audit programs, consumer audit and incentive programs, commercial and industrial audit and incentive programs, landscape regulations and water conservation programs, wastewater management and recycling programs, indoor fixture replacement programs, plumbing regulations, and a water-shortage contingency plan.

Leakage Losses

Tehran's water supply facilities are about 50 years old. According to a leak detection study, the proportion of unaccounted water in the network amounts to approximately 40 percent. This leads to a monetary loss of approximately U.S. \$15 million and a loss of 300 Mm^3 of water per year.

The city of Tehran is tackling leak detection programs on a priority basis. In order to carry out the leakage control work effectively, the Water Bureau tackles such kinds of work as technical development, improvement of instruments, methods, and technical advice to personnel. By means of these progressive leakage-control works, the leakage rate is anticipated to decrease to 20 percent in the next 10 years.

Water losses, or non-revenue water, include physical losses that occur before reaching the consumers (loss in the distribution system), and commercial losses due to inaccurate metering. Under-reading of meters, under-registration of meters, and the tampering with of meters amount to about 7 percent of losses in Tehran.

Breakage and leakages of trunk mains, distribution mains, communication pipes, and illegal connections account for the rest.

CONCLUDING REMARKS

The tremendous speed of population growth in the city of Tehran has raised serious concerns regarding the city's sustainability due to the short supply of water resources and the lack of an integrated approach to its water and wastewater problems. Urban water management in the city will fail without a holistic and integrated view.

Problems are increasing as a result of three factors. First, urban population is growing fast. Megacities like the Tehran Metropolitan Area simply are not able to cope with the growth of the suburbs, and in many cases the development, with its associated increase in water demand, has gotten out of control. The second problem is that because of the lack of adequate water resources and drinking water, water must be transported from far away areas (distances of several hundred kilometers). The third factor is that problems arise with more groundwater exploration and extraction.

The extent of these increasing demands, when placed against the constraints of water, time, funds, and the lack of integrated approaches to manage drought impact and pollution of groundwater, confirms that the city faces a serious water and wastewater crisis.

Curtailment of population growth, water reuse, administrative and technical measures to obtain savings in water consumption, a reduction in losses in the distribution system, and better management of the local water resources are proposed as solutions to Tehran's water problems. These measures are preferred over importing water from outer basins, which results in unnecessary expenditure and ultimately transfers the water problems to other areas.

ACKNOWLEDGMENTS

The authors wish to thank the I. R. Iran and U.S. Academies of Science and Sharif University of Technology for supporting this work.

SELECTED BIBLIOGRAPHY

Tajrishy, M. 1997. Demand management to solve Tehran's problem, Presented at 4th International Conference on Civil Engineering. Tehran, Iran.

Tajrishy, M. Water crisis in Tehran: The integrated approach, Water and Sewage, Volume 22, 1997. Pp. 2-12.

Tajrishy, M. Tehran's wastewater management through a new integrated approach, Water and Sewage, Volume 28, 1998. Pp. 2-11.

Tehran Province Water and Sewage Company. A glance at water and wastewater. Province Company of Tehran, 1998. P. 16.

Jamab Consulting Engineers, National Water Master Plan- Tehran, Water Resources Organization of Iran, 1999.

Mahab Ghodss Consulting Engineering, Optimum Utilization of Groundwater Resources in a Heavily Exploited Aquifer of Tehran, Technical Report, 2001.

Maksimovic, C., and Tejada-Guibert, J.A. (eds.) Frontiers in Urban Water Management. Deadlock or Hope, IWA Publishing, London, 2001.

INTERBASIN WATER TRANSFER

Interbasin Water Transfers in the Western United States: Issues and Lessons

David H. Getches

Sometimes the demand for water justifies transporting it far from the stream where it originates, even into the watershed[1] of another river. Population centers, industrial uses, and some agricultural uses produce enough economic value to compensate for the expense of constructing facilities such as canals, pipelines, and pumping stations. The most typical case occurs when a growing city in an arid region looks to a distant source for a water supply.

Sometimes the source of water to be exported is considered available because it is relatively plentiful and there are low present demands for its use in the watershed of origin. In other instances, water is taken out of established local uses in order to supply new uses in another basin. For instance, cities in the western United States sometimes buy water rights from farmers to meet expanding water demands. This typically results in drying up of irrigated lands.

Economists urge that water transfers should not be impeded by legal restrictions. They argue that it increases efficiency to move water to the most highly valued uses. Yet residents of the areas of origin often resist exporting water. When water is removed from an area it almost invariably causes economic and environmental impacts. These impacts are the greatest in the case of transfers from one basin to another, because once water is diverted there are no return flows so that the water cannot be reused. Several legal measures in the western United States have attempted to ameliorate the impacts on areas of origin with-

[1]In this paper the terms "watershed," "basin," "catchment," and "drainage area" are used interchangeably.

out unduly restricting interbasin transfers of water.[2] These measures range from simply considering the possible impacts, to providing mitigation, to barring transfers. They vary in effectiveness from state to state, and no state has a truly comprehensive approach.

It is difficult to find the appropriate balance between the interests of the area of origin and larger national or state interests. Usually the latter interests prevail, as local areas of origin are usually slower growing and less economically advantaged. Thus, they rarely have sufficient political power to prevent or alter plans to de-water their watersheds. At the same time, there may be significant statewide or national interest to be gained from an interbasin transfer of water, and the area of origin should not have effective veto power over the transfer. Experience in the United States, however, teaches that interests of the area of origin are most typically given little consideration. Indeed, the state or nation may have broader interests in maintaining the ecological integrity or economic viability of a local area, but even these interests have been disregarded in past water development decisions. Most major water development in the United States occurred in an era before the existence of any effective legal controls, including most of those discussed in this paper. This permitted interbasin transfers to proceed relatively unimpeded with little consideration of interests other than those of the importing region.

The case of the now-infamous transfer in California between the Owens Valley and Los Angeles is illustrative. In the 1920s, the city of Los Angeles secretly purchased the water rights of farmers in the Owens Valley, some 200 miles from the city. By the time the public discovered that the valley had lost all of its water and the farms and associated economy would be dried up, it was too late to stop the huge interbasin transfer. The city built a huge pipeline to carry the water to Los Angeles residents. The people of the Owens Valley protested and forcibly tried to keep the pipeline from operating. They were unsuccessful, although protests, demonstrations, and occasional dynamiting of the pipeline continued intermittently for 80 years.

Only after the state began to legally recognize the importance of environmental interests, including a requirement of environmental assessment, and the state courts of California recognized that the state holds all water in a public trust, did the people of Owens Valley attain any vindication of their rights. The Inyo County government and environmental groups brought legal actions in several courts. After years of litigation, the city of Los Angeles has now been forced to reduce its diversions, to mitigate some of the environmental damage,

[2]In the United States, Indian tribes have sovereign powers to regulate resources within their territory. Many of the actions that states can take to control or mitigate the impacts of exports can be considered by tribal governments as well as by states. The nature and extent of tribal sovereignty are not discussed here, nor are the special issues that tribes confront concerning proposals to move water away from their reservations.

and to recognize the interests of the county and its residents. On their own the residents of Owens Valley lacked sufficient political influence, but after environmental protection has become a legally recognized state interest, the valley's concerns could be at least partially vindicated.

Although urbanization of the West surely will result in calls for more interbasin transfers, opportunities to develop such projects are limited by extensive past development. Most of the West's rivers are fully developed. Nevertheless, the Owens Valley example illustrates that relatively new legal innovations for protecting the interests of areas of origin may be important in ameliorating the existing diversions. As existing facilities are put to new uses, rehabilitated, or expanded, protections for areas of origin enacted in recent years or in the future may apply to these changes as well as to new developments.

PRIOR APPROPRIATION PRINCIPLES

To understand the framework for the efforts that have been used to deal with interbasin transfers in the United States, it is useful to have a basic understanding of the system of water rights that prevails in the West. The law creates no fundamental barrier to taking water from one watershed to another. In the prior appropriation system, unlike the riparian system water rights that prevails in the eastern United States, water is not legally tied to the land or to the watershed.

Prior appropriation began as a simple system that allowed private water users to obtain a water right recognized by the state by diverting water out of a stream and putting it to a "beneficial use." At first, it was not necessary to ask permission but only to "appropriate" the water. The water right obtained in this way entitled water users to continue taking the full amount of water that they originally had diverted. But the water users would lose the right if they did not continue using it ("use it or lose it"). When there was insufficient water in the stream, users were not required to share the resources. The users with the oldest rights were entitled to take the full quantity of their rights. Once the most senior users exhausted the available water, all water rights holders junior to them would be cut off entirely ("first in time, first in right"). Once a person held a water right, it generally could be put to other uses, used on other lands, and transferred to others, so long as the change in use did not harm the water rights of any other person. Because water rights were considered a form of property, these rights could be transferred to others.

Today, the system of water rights in most western states is still guided by the basic principles of the prior appropriation doctrine, but the rules have been codified and expanded in statutes. Moreover, rights are allocated and administered by state agencies and officials who maintain records, enforce rights by opening and shutting headgates and dams, and make decisions that apply statutory criteria. When someone applies for a new water right to change the use of an existing right, most state laws require a finding by a state official or agency that granting the application will be consistent with "the public interest" or "public welfare."

Some of the earliest cases in which the courts recognized the prior appropriation doctrine involved distant transfers, even to other drainage basins (*Coffin v. Left Hand Ditch Co.*, 6 Colo. 443 [1882]). The riparian doctrine that was applied in many eastern states would not allow transbasin diversions if it were strictly applied, and so the courts in the West decided that they would follow the "first in time, first in right" custom of the early miners, which applied to water use as well as to mineral extraction. This system did not restrict movement of water from a stream to wherever it was needed. In the arid West streams were few and flows fluctuated greatly, so the law was designed to accommodate the needs of those who first arrived and had economic uses for the water. Judges in the leading cases rejected riparian principles and embraced the prior appropriation doctrine based on the assumption that the latter was better suited to the geography and climate of the region.

THE DEMAND FOR INTERBASIN TRANSFERS

When water users in one basin want to use water from another basin, they may apply for rights to water that is not being used by others, or they may purchase rights to water that is already in use in the basin of origin. On most streams in the West all the available water had been typically appropriated more than one hundred years ago so that new interbasin transfers must result from transfers of existing rights. In less developed areas, however, there still may be unappropriated water available for new rights.

It costs nothing to initiate a new appropriation, but it is expensive to purchase a senior water right, because the priority of an older right makes it more secure in times of shortage and therefore more valuable. Many water users like municipalities are willing to pay the additional expense to acquire older water rights, because even in a watershed that is not fully developed, the senior right will provide a more reliable supply of water. Purchasers may also choose to buy senior water rights from far away rather than local senior water rights, although they must incur substantial costs for transporting the water. Usually senior rights close to a developed area are much more expensive than senior rights in a distant basin, and the difference in price may be greater than the cost of building tunnels and pipelines. But cost is not the only factor in a decision to import water. Sometimes cities will choose to pay the cost of transporting imported water from a different watershed to avoid the political controversy of removing water from farms near them.

Today, transfers of water rights are an important part of water law and administration. Water rights can be transferred for payments of money, and increasingly we speak of water markets. Marketing enables water to be moved out of old uses that have become less productive, such as mining after the minerals have been removed or agriculture in areas that produce low-valued crops. The most frequent purchaser is a city that needs a larger water supply to serve a growing population. Indeed, urbanization is a major trend in the West. The fastest growing new demands for water in the West are in cities that can pay farmers enough to induce them to sell because of the relatively low economic benefits from agricultural water use. Cities also are able to pay the costs of building facilities to transport and store water, because they can spread the cost among large numbers of users, and their budgets are sometimes subsidized by general tax revenues.

EFFECTS OF INTERBASIN TRANSFERS

Diverting water from an area can cause a variety of negative economic, social, and environmental consequences. When water supplies are committed to exports, it discourages investments in new businesses that require water. The transfer of water to another watershed therefore deprives the area of origin of water that could be used for economic growth in that area of origin. It would be ironic for a water-rich area to seek water somewhere else because its local supplies are being exported. The option of "buying back" the water that is being exported exists, but the price may be too high.

The impacts of water exports are more palpable when the water being transferred is already being used in the area of origin. The seller of the water rights—such as a farmer selling irrigation rights—presumably will be paid the fair market value of the rights. Although the seller receiving compensation will not suffer hardship, third parties may suffer indirect but significant economic impacts. As the farming economy declines, so will the businesses that depend on selling tractors, seeds, and fertilizer and the banks that lend money. All the businesses that depend on these businesses are, in turn, affected. With less business activity, local governments will collect less tax revenue, causing a decline in the ability of local governments and school districts to provide services to citizens. As community life declines the area will becomes less attractive to new businesses resulting in a downward spiral of economic effects.

Almost invariably, interbasin transfers cause environmental impacts. As streamflows are diminished, fish and wildlife habitat can be diminished and destroyed. Wetlands may dry up, and native vegetation may die out. If lands were previously being irrigated and water is then transferred away, the fields can turn to dust and the soils can blow away, degrading the arability of the lands as well as causing pollution problems. Municipalities that are required by law to maintain a certain water quality may find that reduced streamflow will increase their costs of treating sewage, because it is more difficult to dissolve the discharged pollutants in order to meet water quality standards.

Environmental problems also can adversely affect the area's economy. Many rural areas of the West are becoming increasingly dependent on tourism. Tourists will not come if they are seeking an unspoiled natural environment, including free-flowing streams that are valued for fishing, hunting, boating, and aesthetics. Such areas also may become unattractive to new businesses who are seeking a location with a pleasant, healthy environment.

The history of the development of the Colorado River in the southwestern United States illustrates many of the problems of interbasin transfers and the historical inadequacy of government responses. The Colorado River is the major river for the large, arid southwestern region. Much of the region within the watershed of the Colorado is rural and sparsely populated. It is also an area of great natural beauty, with wildlife, fish, and recreational opportunities. The watershed of the river includes parts of seven states and Mexico. The states have fought over allocation of water, with some of their battles being resolved by interstate agreements known as compacts, and others being mediated by the Supreme Court and Congress. Disputes between the United States and Mexico have arisen periodically and have been narrowed by the negotiation of a treaty and subsequent amendments.[3]

Almost all of the water of the Colorado River is exported to other watersheds for use in growing cities as well as on farms. In the state of Colorado ten tunnels have been drilled through the Rocky Mountains to carry the water away from the basin of origin to population centers east of the Continental Divide. Similarly, in Utah water is taken hundreds of miles to cities on the west side of the Uintah Mountains. California transports Colorado River water through huge aqueducts to farming areas and to large cities on the west coast, including Los Angeles and San Diego. A huge share of the water also goes to California's Imperial Valley near the Mexican border, mostly for large farms owned by wealthy families and corporations. Arizona pumps water out of the river and up 1500 feet so that it can flow hundreds of miles across the desert through a large canal. A series of dams on the river stores water to feed these out-of-basin uses. Most of the dams, canals, pumping plants, and tunnels were constructed with federal funds.

By the time the Colorado River reaches Mexico, in most years there is barely enough flow remaining to deliver water that was promised to Mexico in a 1944 treaty. That water, too, is taken mostly for uses away from the river. After that, the river has been entirely depleted in all but very wet years. The delta in Mexico has changed from an area rich in fish, birds, and wildlife to a dusty plain with a trickle of water occasionally threading through it.

Interbasin transfers from the Colorado River have profoundly affected ecosystems, communities, and economies and have been the cause of hostilities and

[3]Treaty between the United States of America and Mexico relating to the utilization of the Waters of the Colorado and Tijuana Rivers and of the Rio Grande signed at Washington February 1944; protocol signed at Washington November 14,1944, Entered into force November 8,1945, 59 Stat. 1219; Treaty Series 994.

dissension between the people in the watershed and those in the areas where the water is used. Some of the problems caused by the transbasin diversion of virtually all of the Colorado River are as follows:

- Nearly all species of native fish in the river are either extinct or endangered.
- Depletion of flows has allowed salts to concentrate in the river so that at times the water is so saline that it kills crops.
- Dozens of communities within the natural watershed of the Colorado have limited options for future growth, because control over the use of the water that originates there has been legally vested in large cities and districts that export the river's water to other watersheds.
- The Colorado River Delta has lost most of its native species of birds and fish, its wetlands are in danger, and its indigenous communities have nearly been destroyed.

ORIGIN PROTECTION LAWS

Because of the economic benefits of allowing a resource like water to be moved to uses that have the greatest value; it is theoretically undesirable to inhibit such transfers. Economic theory, however, assumes that all costs are accounted for in determining that there is a net benefit to society. There are two problems with this. One is that the impacts on third parties are not, in fact, considered by the parties to the transaction and, in any event, they are difficult to quantify. Thus, these costs are "externalities" that escape the calculus. The other problem is that while the nation's economy may be better off if the farmers and the businesses that depend on them in a local area go out of business and a new city grows, there are nevertheless major negative consequences for the local community. It is politically, and perhaps morally, unacceptable to ignore these consequences.

Although public policy does not favor prohibiting or creating major impediments to transfers, most experts recognize that basins of origin have valid interests that the law should not ignore. Nevertheless, state laws and policies often fail to recognize the economic, social, and environmental harms that can occur when water is removed from its natural watershed. Prior appropriation law has never limited the use of water to the watershed in which it originates. Consequently, efforts to deal with the problems of interbasin transfers sometimes encounter objections that these efforts are fundamentally at odds with western water law that has placed no limits on where water could be used.

LEGAL MEASURES DIRECTLY CONTROLLING INTERBASIN TRANSFERS

In 1973, the National Water Commission's Report recognized the importance of the issue and recommended that transfers from one basin to another be allowed only when interbasin water transfer provides the lowest cost source of water and the benefits exceed the costs. No state has adopted any such law or policy. States, however, have implemented a number of legal approaches that provide protection for the interests of basins of origin. Some legislation directly controls interbasin transfers. More commonly, laws operate indirectly to limit or mitigate the effects of interbasin transfers. We first consider direct types of controls on interbasin transfers.

Prohibition or Severe Restriction

In the past, some state laws made certain types of transfers unlawful and made others rather difficult. A few states made water rights appurtenant to specific lands and flatly prohibited their transfer. As early as 1909, Wyoming passed a statute that prevented transfers of water out of agricultural uses. An Arizona law makes transfers of water beyond the boundaries of irrigation districts subject to the approval of the district (Arizona Revised Statutes § 45-172[5]). This effectively gives basin of origin interests a veto power over transfers. As such, it disregards any broader interests a state may have in providing water for economic expansion and human needs in the importing area.

Compensatory Storage

A Colorado statute attempts to protect the area of origin by requiring that any conservancy district proposing to export water from western Colorado must ensure that present and future uses of water are not impaired nor increased in cost by the proposed project.[4] The law was interpreted by the Colorado Supreme Court in a case involving the Windy Gap Project that would take water from the Colorado River through a tunnel under the Rocky Mountains to cities in the urbanized eastern part of the state. The court required the importer to construct storage facilities in the area of origin (Colorado River basin) and held that the importer may not proceed without presenting well developed plans for the reser-

[4]"(IV) However, any works or facilities planned and designed for the exportation of water from the natural basin of the Colorado River and its tributaries in Colorado, by any district created under this article, shall be subject to the provisions of the Colorado River compact and the 'Boulder Canyon Project Act.' Any such works or facilities shall be designed, constructed and operated in such manner that the present appropriations of water, and in addition thereto prospective uses of water for irrigation and other beneficial consumptive use purposes, including consumptive uses for domestic, mining, and industrial purposes, within the natural basin of the Colorado River in the state of Colorado, from which water is exported, will not be impaired nor increased in cost at the expense of the water users within the natural basin. The facilities and other means for the accomplishment of said purpose shall be incorporated in and made a part of any project plans for the exportation of water from said natural basin in Colorado." *Colorado River Water Conservation Dist. v. Municipal Subdistrict*, Northern Colorado Water Conservancy Dist., 198 Colo. 352, 610 P.2d 81 (Colo. 1979).

voir before the proposed diversion for the Windy Gap project could be approved. Under this law the importer must present detailed plans for the compensatory storage facilities and demonstrate that they will be adequate before the water court will approve a diversion of water over the Continental Divide (*Colorado River Water Conservation Dist. v. Municipal Subdistrict*, Northern Colorado Water Conservancy Dist., 198 Colo. 352, 610 P. 2d 81 [Colo. 1979]).

It is interesting that, although the district eventually developed the detailed plans required by the court, the plans were never used. First, the importing district agreed to pay up to $10 million to construct a compensatory storage reservoir known as the Azure Reservoir. Some years later, the parties agreed that the Azure Project should no longer be pursued. Instead, the importing district agreed to pay $10.2 million to the water district in the area of origin so that it could build another storage project of its choosing. The water court approved.

The district that proposed the Windy Gap Project made other concessions in order to resolve objections of water users and others. Although no specific statute required it, the district granted the use of storage space to a western slope water district in an existing reservoir owned by the importing district but located in the basin of origin. It also agreed to subordinate the use of its water rights to users in the basin of origin. Payments were also made to upgrade water treatment facilities and to study salinity problems.

The concept of compensatory storage embodied in the Northern Colorado Conservancy District Act (May 13, 1937) is based on an accord reached a few years before the Act was passed. The agreement was necessary to reconcile eastern and western slope interests that were to be affected by the Colorado–Big Thompson (CBT) Project. The CBT Project was planned to take water from the upper reaches of the Colorado River and transport it across the Rocky Mountains to the northern Front Range. Negotiations resulted in a resolution that included several principles for transmountain diversion projects. One was that an "essential part" of such projects should be the "construction of a compensatory reservoir on the western slope of sufficient capacity to hold an amount of water equal to the amount to be annually diverted . . ." This and other principles were included in a legislative report called "Senate Document No. 80, (75th Cong., 2nd session, June 24, 1937)" which in turn was incorporated by reference in the CBT authorizing legislation. Accordingly, the 152,000 acre-foot Green Mountain Reservoir was built on the western slope as part of the CBT.

The Bureau of Reclamation is still searching for purchasers of Green Mountain water; some fifty years after the reservoir's construction, the water remains virtually unused except for electric power generation. It is unsurprising that the National Water Commission's 1973 Report was critical of compensatory storage schemes in that they may cause "economic waste because the area of origin may not be prepared to use the compensatory storage for many years" (National Water Commission, 1973).

The compensatory storage requirement specifically applies only to conservancy districts and only to exports from the basin of the Colorado River. Therefore it does not apply to the larger and more numerous transbasin diversions in Colorado that have been undertaken by Denver and other municipalities. Colorado law provides no area of origin protection against the effects of those transfers.

Priority for the Area of Origin

States may grant the area of origin a priority right to appropriate the water it may need in the future. The importing basin technically can use water only until the basin of origin needs the water, at which point the basin of origin can recapture its water and invalidate transfer contracts. If enforced, this right obviously would lead to unsettling results for the importer of water. More likely as a practical matter, the importing area could use political and economic power to prevent the area of origin from interrupting long-standing water deliveries in the importing area.

California's area of origin protection laws ostensibly give the exporting area absolute priority to make future use of water over that of the importing area. California's population is concentrated far from most water supplies. Thus, growing cities where water demand is high depend on removing huge quantities of water from sparsely populated areas with copious water.

One California statute reserves for the county of origin all the water it may need for future development (Cal. Water Code §§ 10505–10505.5 [1992]).[5] The California law provides that the State Water Resources Control Board makes the determination of when, and to what extent, water is "necessary for the development of the county." This board is a quasi-judicial body set up to regulate water rights, water pollution, and water quality, whose five members are appointed by the Governor. It is easy to imagine a dispute arising between, say, the Metropolitan Water District of Southern California (the main recipient of State Water Project water) and a small county in the north as to the amount of water that is "necessary for the development of the county." Once water from the area of origin has been committed to use in another area, the importing area is likely to grow dependent on the water supply. Moreover, investments and plans for growth in

[5] § 10505. Restrictions on release or assignment, No priority under this part shall be released nor assignment made of any application that will, in the judgment of the board, deprive the county in which the water covered by the application originates of any such water necessary for the development of the county.

§ 10505.5 Territorial restrictions on use, Every application heretofore or hereafter made and filed pursuant to Section 10500, and held by the State Water Resources Control Board, shall be amended to provide, and any permit hereafter issued pursuant to such application, and any license issued pursuant to such a permit, shall provide, that the application, permit, or license shall not authorize the use of any water outside of the county of origin which is necessary for the development of the county.

the area of origin may be restrained by the fact that water is committed to the importing area, even if there is a legal right to demand the water for future growth in the area of origin.

California's county of origin statute protects only those counties in which the water "originates." This wording has been construed to mean that counties are protected only to the extent that water actually originates, in the form of rain or snow, within their boundaries. This interpretation raises difficult problems for the foothill counties. In California, these counties usually receive minimal rainfall but have abundant water in the rivers that are fed by mountain runoff and snow melt. Thus, most of the water "originates" in mountainous counties that are usually too rugged for much future development. The foothill counties downstream, whose potential for development is often much greater, received little protection from the county of origin law for the abundant water naturally flowing through them. This problem became urgent when the state decided to construct the massive Central Valley Project. To solve it, the legislature enacted another area of origin protection law known as the Watershed Protection Act. This was a part of the Central Valley Act of 1933 (Cal. Water Code §§ 11460–11463[1992]).[6]

The main idea of the Watershed Protection Act was to extend area of origin priorities to the entire watershed area and not limit them to the areas of precipitation. This grants some relief to the foothill areas, which are almost always included in the region or area that contributes to the supply of the stream in question (25 Ops. Cal. Atty. Gen. 8, 19 [1955]). A proviso that the "article shall not be so construed as to create any new property rights . . ." precludes the board from assigning water rights to an area based on a determination of its potential need. Instead, water users in the area have a priority that allows them to recapture water from any other non-watershed user when they require it in the future.

The Watershed Protection Act was enacted in 1933 and was meant to apply to the state's Central Valley Project (CVP). The Depression was, however, well underway and the state was unable to finance the CVP. The federal government took over the project in 1935 and by 1940 was deeply involved in the water

---

[6]§ 11460. Prior right to watershed water.

In the construction and operation by the department of any project under the provisions of this part a watershed or area wherein water originates, or an area immediately adjacent thereto which can conveniently be supplied with water therefrom, shall not be deprived by the department directly or indirectly of the prior right to all of the water reasonably required to adequately supply the beneficial needs of the watershed, area, or any of the inhabitants or property owners therein.

In no other way than by purchase or otherwise as provided in this part shall water rights of a watershed, area, or the inhabitants be impaired or curtailed by the department, but the provisions of this article shall be strictly limited to the acts and proceedings of the department, as such, and shall not apply to any persons or state agencies.

§ 11462. Creation of new property rights

business in California. It was nevertheless assumed that the county of origin act and the Watershed Protection Act were still effective since § 8 of the federal Reclamation law purported on its face to incorporate state law to control the way water rights are acquired. Recently, a major agricultural water district has taken legal action under the statute asserting that it has rights to water that originates in its area but was developed under the CVP and used for many years by water districts outside the area (California Environmental Insider, 2000).

Public Interest Considerations

Most states legally recognize a strong public interest in water. State constitutions and statutes make water a public resource to be held in common for all citizens until private rights are established in it. Therefore, water rights allocation and transfer must be consistent with the public interest or public welfare. In practice, states rarely deny new uses or transfers in order to protect the public interest, but they do impose additional conditions on the water use.

The diverse economic, social, and environmental impacts of interbasin transfers are typical of the concerns often expressed under the banner of the public interest—instream flows, fish and wildlife, recreation, water quality, economic viability of an area, and others. Yet, few state laws clearly indicate what factors are to be considered as within the public interest.

State law requires Idaho's Department of Water Resources director to determine whether a proposed water use is in conflict with "the local public interest," but the statute does not articulate what this means (Idaho Statutes, sec. 42-203A). Therefore, the Idaho Supreme Court has read the statute with reference to other Idaho laws and the laws of other states that use similar terminology (*Shokal v. Dunn*, 707 P.2d 441 [1985]). Since that decision, the Director of Water Resources has convened hearings aimed at reaching decisions that ensure "the greatest benefit possible to the public [from public waters] for the public" (*Shokal v. Dunn*, 707 P.2d at 448 (citing *Young & Norton v. Hinderlider*, 15 N.M. 666, 110 P. 1045, 1050 [N.M. 1910]). Affected citizens can present evidence about matters such as aesthetics, recreation, fish, and ecosystem functions that will be impacted by the proposed water decision. The agency considers not only benefits to the applicant but also economic effects, alternative uses, minimum stream flows, wastewater, and conservation.

Modern permitting systems attempt to limit or prevent adverse impacts on the public from water uses at the time new users obtain permits. But not all states apply their public interest requirements to changes of use or transfers. The Supreme Court of Utah, however, upheld the application of the same criteria to changes in use that it applies to new appropriations (*Bonham v. Morgan*, 788 P.2d 497 [1989]). In Nevada, a statute requires the state to reject an application for a water transfer that would result in damaging the public interest (Nevada

Revised Statutes § 533.370[3]). Wyoming, one of the few states with a special process to evaluate transfers, considers potential economic losses to the community relative to the benefits of the transfer and the availability of other sources of water (Wyoming Statutes Annotated § 41-4-503). California, through the State Water Resources Control Board, reviews proposed transfers to determine if they would cause an unreasonable effect on the economy in the area of origin or on fish, wildlife, or other water uses (California Water Code § 109).

Although every state in the West except Colorado uses some type of process to review the public interest in water decisions, they all could improve the way in which they apply these laws. The majority of the states lack clear standards to define the public interest that they are trying to protect. Many of the social, economic, and ecological interests affected by water allocation, transfer, and use are simply not included in the considerations of state agencies. If the elements constituting the public interest were comprehensively articulated, government employees could use them to guide state policy in resolving conflicts among competing interests and to understand better the tradeoffs inherent in any water decision.

Another problem is that the public does not always have a role in the process used to review public interest compliance. The processes used to issue permits in some states, for instance, involve only water rights holders and not the members of the public who experience economic, environmental, and social impacts from water use and development. One way that states can consider the public interest in water and include the affected public is through comprehensive water planning processes that articulate the kinds of interests the public has in particular watersheds and that enunciate state policies related to them. The plans can then guide the application of public interest laws.

Public Trust Doctrine

The examples described above require public interest review before a new water use or changed water use is approved. In some instances, however, courts have held that a state's decision to permit private use of public resources can be voided when water rights are allocated or transferred without review of the public interest (*National Audubon Society v. Superior Court,* 658 P.2d 709 [Cal. 1983]; *In re Water Use Permit Applications,* 9 P.3d 409 [Hawaii 2000]). The public trust doctrine recognizes that water is fundamentally a public resource and that the state should not allow private water rights to impair the public's interest in water. As applied, the doctrine allows a court to reexamine established water rights in order to ensure that environmental values are not destroyed without prior consideration of the impacts. The doctrine has its origins in civil and common law principles that recognize the public's continuing rights to use navigable waters and the state's property rights in the beds of navigable waters.

Instream Flow Protection

Laws creating state programs to protect instream flows indirectly address significant issues that arise from interbasin transfers. Most of these laws allow state agencies to enforce minimum streamflows. The state itself may be empowered to appropriate rights to quantities of flowing water, or it may be authorized to reserve an amount of water that is necessary to maintain desired flows from the water that is available for appropriation by private parties. At present, only Arizona and Alaska permit entities other than the state to appropriate waters for instream flows. In all other states only a state agency may hold the right.

The recently enacted statutory programs that allow states to appropriate water rights to protect streamflows can be criticized, because the water rights in most streams usually have been appropriated before the laws were passed. Thus, it is possible for senior rights to dry up the stream most of the time. In these states, effective protection of the streamflows ultimately will depend on the acquisition of senior water rights. Some states do permit the state to buy or accept donations of senior water rights with priorities sufficient to maintain streamflows all or most of the time. Private groups have formed "water trusts" to finance purchases of senior water rights; these rights must be transferred to the state agency authorized to hold instream flow rights unless the state allows private entities to hold them.

Instream flow protection laws do not protect water for some public uses. The Colorado statute permits appropriations of a quantity of water sufficient "to protect the natural environment to a reasonable degree" (Colorado Revised Statutes Annotated § 37-92-102[3]). The state board that holds the rights has interpreted this language narrowly and has used it almost entirely to protect coldwater fish such as trout. So applied, the law is generally unavailable to protect water quality, riparian vegetation, wetlands, or recreation.

Reservation of Rights for Future Use

Montana has a system for reserving future rights for the future needs of municipalities. The reservation of a water right protects against all future appropriations, not just transbasin exports, that would interfere with future municipal uses. Reserving a specific quantity of water for the future requires "informed guessing" in order to project future demand. These guesses will invariably be wrong. Thus, the area may end up with too little water (at least from its perspective) and no chance of acquiring more because all the existing supplies are subject to long-term commitments. Or, the area may end up with a claim to more water than it needs that discourages others outside the area from developing and using the water.

Land Use Regulation

Colorado has an unusual law that allows local governments to regulate land uses that are of statewide significance (Colo. Rev. Stat. § 24-65.1-101). This generally applies to very large developments or those that affect a large area. It includes large water development projects. By using this law, some counties have prevented major water collection facilities and pipelines from being built in their territory to take water to another watershed (*City & County of Denver v. Board of County Comm'rs*, 760 P.2d 656 [Colo. Ct. App. 1988], aff'd, 782 P.2d 753 [Colo. 1989]). These laws give extraordinary power to the area of origin to protect its interests, although they can be criticized as disregarding broader interests of the state in allowing water to be used where it is needed the most. On the other hand, local officials in the area of origin—often susceptible to schemes that provide short-term economic benefits—may be persuaded to compromise environmental and other public interests and allow a destructive project to proceed.

Federal Environmental Laws

Environmental laws that protect water quality, wetlands, and endangered species indirectly deal with the effects of interbasin transfers. If federal permits are required, the permitting authority can insist that the project be modified to reduce the environmental impacts or that mitigation measures be used. These laws can prevent particularly damaging interbasin transfers from proceeding. More likely, they result in negotiated changes in the proposed transfers that mitigate the harm.

Nearly all environmental laws are federal, and therefore federalism concerns arise when environmental laws conflict with or curtail the uses of water under state water rights. State laws are generally not very strong or effective in protecting the public's interest in water, and some water laws say that if rights to use water are in conflict with environmental laws, water rights will prevail. Yet the Constitution makes federal law supreme.

The National Environmental Policy Act of 1969 (NEPA) (42 U.S.C.A. §§ 4321-4370, 4321[2][a][1]) requires federal agencies to assess potential environmental impacts of proposed "major federal actions." The agencies must hold hearings that allow public participation and then prepare a document known as an environmental impact statement (42 U.S.C.A. § 4332[2][c]). NEPA applies to proposals that require a federal approval or license or that will use water from a federal water project if there will be a significant environmental impact. A few western states, including California and Washington, have adopted laws with similar environmental assessment requirements for projects permitted or sponsored by the state. The state or federal laws that require an assessment of environmental impacts are important mechanisms for evaluating the effects of water development and transfer. NEPA is essentially a procedural requirement and does not mandate that a final decision be environmentally benign. It only requires that the agency adequately present complete information before making its decision.

The federal Clean Water Act (CWA) protects water quality (33 U.S.C.A. §§ 1271-1387). Although it is a federal law, the CWA is actually administered by most states. Under the CWA, anyone who makes a "point source" discharge of pollutants (i.e., a discharge from a pipe or ditch) into the waters of the United States must have a permit that limits the quantity of particular pollutants according to standards established by the federal government (33 U.S.C.A. §1362[14]). The permit also must require sufficient limitations on discharges to protect the overall quality of the watercourse receiving the wastewater. Standards for water quality are set by the states and are specific to particular waterways. The permitting program under the CWA has effectively regulated industries and municipal sewage treatment plants that discharge wastes into rivers and lakes.

The CWA does not deal with declines in water quality caused by other than point source discharges. Yet, when a stream is depleted, especially by an interbasin transfer that has no return flow to the basin of origin, water quality declines because any waste added naturally or by humans to the stream becomes more concentrated. There are, however, no formal controls of water depletions that damage water quality.

A special program under § 404 of the Clean Water Act regulates dredging and filling of navigable waters. The statute defines navigable waters as all "waters of the United States." This has been interpreted administratively to include all adjacent wetlands, and wetlands are defined as any area capable of sustaining riparian vegetation. The activities covered extend beyond the dredge and fill operations undertaken to deepen channels for navigation. Depositing fill material can include any construction in a waterway or wetland. Thus, the statute covers water projects, dams, and diversion structures.

The impact of § 404 on interbasin transfers is much greater than this description might suggest because almost any type of construction activity to develop or use water—and certainly nearly every interbasin transfer from a stream—involves facilities that are on the banks of a stream or that are in or cross "wetlands." Where these areas are affected by water development activity, § 404 requires the United States Army Corps of Engineers to conduct a global review of the public interest. This gives the Corps extensive authority to review the impacts and require mitigation for the impacts of a water development on the basin of origin.

The Endangered Species Act (ESA) (16 U.S.C.A. §§ 1531-1543) is another federal statute that can affect proposals to divert, develop, or transfer water. The ESA absolutely prohibits any action by the federal government that would jeopardize the continued existence of an endangered species. Federal agencies considering activities that could have this effect are required by § 7 of the ESA to consult with the U.S. Fish and Wildlife Service. If, in the opinion of that agency the action would jeopardize the endangered species, the action cannot proceed unless there is a reasonable and prudent alternative that will not cause the jeopardy.

The ESA is extremely powerful because nearly every major water project—not just those undertaken directly by the federal government—either requires some kind of federal approval (such as under § 404 of the Clean Water Act), or receives federal financing. Thus, the ESA has proved to be a formidable barrier to water development that could be destructive of fish or wildlife habitat where endangered species are found. The Act, indeed, may be the most significant law affecting new water development. It protects any affected species in the area of origin, the importing area, and in the area of any pipelines or other facilities.

Another section of the ESA, § 9, prohibits actions that "take" or "harass" an endangered species. These terms are broadly interpreted to include even harm to the habitats of endangered species. Unlike § 7, which is specific to federal agency actions, § 9 extends to private actions. The section has rarely been applied to restrict private water development or uses. In one exceptional case, however, the pump diversion facility of an irrigation district killed several endangered salmon. A federal court enjoined the irrigation district's activities, prohibiting it from "taking" the endangered species (*Department of Fish and Game v. Anderson-Cottonwood Irrigation District,* 11 Cal.Rptr.2d 222 [Cal. App. 1992]).

Public Participation Laws

Communities, local governments, individuals, and nongovernmental organizations dedicated to environmental protection often demand to be heard when water is transferred from one area to another. Environmental interests had no place in influencing water decisions under the early laws of the West, and even today the interests of the environment in the areas from which water is proposed to be exported are not adequately represented in the laws and policies of many western states.

Many of the laws that were discussed above require public hearings or processes that can be forums for members of the public to present their views to government decision makers. Although water allocation decisions once were made by the state with input only from people who held water rights, today the decision making processes of many states have been opened up to the public. Public agencies usually receive comments from parties directly involved in decisions concerning the allocation, development, and transfer of water rights. Modern laws have begun to include various interests including third parties affected by the allocation of water rights and other decisions concerning water use. All of the federal environmental laws also have public interest participation requirements.

Collaborative Efforts

Stakeholders (water rights holders, local residents, businesses, environmental groups, and others) who are affected by a water development or transfer have begun to organize and to exert substantial influence over water decisions, not only those that involve interbasin transfers of water. Dozens of initiatives throughout the western United States demonstrate the potential for these locally based, problem-solving entities.

In some cases, such groups have solved problems of diminished streamflows in popular fisheries by crafting voluntary agreements among water rights holders to change the timing of withdrawals. Sometimes third parties affected by a proposed interbasin transfer are able to persuade the proponents to take voluntary action to protect the interests of the public. Unless there is a public process provided for under the law, it is difficult to initiate these negotiations. Only when third parties have sufficient political or legal leverage (for example, the threat that a federal law like the Endangered Species Act will prevent their project from being built) will the proponent of the development activity participate in negotiations. When there is no such legal threat it is difficult for third parties with little political power to get the attention of people who want to develop water. The results, then, are not consistent among similar projects and often provide incomplete relief where the objectors lack political or legal strength.

CONCLUSION

The western states have made some progress in addressing the impacts of interbasin transfers; however, their efforts are incomplete. Although it can be argued that the most damaging interbasin transfers have already been constructed and most of the damage has been done, pressures will continue to move water out of more slowly growing watersheds to expanding urban areas. Therefore, it is wise to consider changes in state laws that will prevent the kinds of harms that have been experienced in places like the Owens Valley and the Colorado River basin.

A committee of the National Academy of Sciences (NAS) studied the effects of water transfers in the West. The committee's report (NRC, 1992) made the following recommendations that apply to interbasin transfers:

- States and tribal governments should develop specific policies to guide water transfer approval processes regarding the community and environmental consequences of transferring water from one basin to another, because such transfers may have serious long-term consequences.
- Water transfer processes should formally recognize interests within basins of origin that are of statewide and regional importance, and these interests should be weighed when transbasin exports are being considered.
- Although each state or tribe should select the approach that suits its needs best, protection of areas of origin generally would include impact assessment, opportunities for all affected interests to be heard, regulatory mechanisms to help avoid adverse effects, compensation (e.g., financial payments or mitigation), and authority to deny a proposed transfer or water use involving a transbasin export if the effects are judged unacceptable.

• States should revise laws that now exempt water facilities from taxation by the county of origin either because the exporter is a public entity or because of provisions that make such facilities taxable only in the county where the water is used. Mechanisms to compensate communities for transfer-related losses of tax base, such as an annual payment in lieu of taxes, may be needed.[7]

Since publication of the NAS committee's report in 1992, states have not made any major improvements in their laws affecting interbasin transfers. The recommendations nevertheless provide useful criteria for evaluating and developing approaches for dealing with the consequences of moving water from one watershed to another. Analyzed in this light, the greatest inadequacies of present approaches in the United States are as follows:

• Decisions do not rely on policies or plans that have been formulated in advance to articulate clear public interest considerations for communities and the environment.

• There are no methods for weighing the relative importance of national, regional, and local interests when water is removed from the basin of origin.

• Although several legal devices, mostly operating indirectly, assess or regulate the effects of interbasin transfers, and some water laws include consideration of the public interest, schemes to provide comprehensive mitigation and compensation to the basin of origin have not been tried.

The states should enact legislation to deal with all of these issues. Laws presumably will apply to modifications of existing interbasin transfer facilities as well as future diversion. Other countries in arid and semiarid regions almost certainly will be confronted with the prospect of interbasin transfers. The principles expressed here can be considered for application there to the extent they fit the legal, social, and environmental situations of those countries.

REFERENCES

California Environmental Insider. Westlands rights application threatens to ignite valley water war. August 31, 2000.

National Research Council. 1992. Water Transfers in the West: Efficiency, Equity, and the Environment. Washington, DC: National Academy Press.

National Water Commission. 1973. Water policies for the future. Final report of the commission. Subcommittee on Water Resources and Environment, Committee on Transportation and Infrastructure, U.S. House of Representatives.

[7] National Research Council, National Academy of Sciences, Water Transfers in the West: Efficiency, Equity, and the Environment, 257–59 (1992).

Interbasin Water Transfers in Iran

Ahmad Abrishamchi and Massoud Tajrishy

ABSTRACT

Iran is an arid and semiarid country with scarce and sensitive water resources. The increasing demand for water has caused an alarming decrease in annual per capita renewable water resources, currently estimated to be about 2,000 m^3 for a population of around 65 million. With the current trend in population growth, the per capita water available is predicted to decrease to less than 1,000 m^3 by the year 2025, which will put Iran in the category of countries with chronic water scarcity. As available water resources are unevenly distributed in terms of both time and space, water resources in many areas are already under pressure.

Over the past two decades, much effort has been expended and great strides have been made in the development and exploitation of water resources for various uses. Primarily these efforts have been aimed at maximizing water supply to meet increasing demands. At the same time, enough attention has not been paid to integrating other approaches in which emphasis is placed on managing demand and saving water through sound water consumption practices.

The uneven distribution of water across the country, on the one hand, and the expansion of population centers as well as developments in agricultural and industrial activities, on the other, has led to the present shortage of water for urban and domestic uses. Acting upon a traditional approach to water management, water managers at both local and national levels find water transfer from humid zones to arid zones to be the only option to satisfy demand. Each year proposals are presented to the government for new interbasin water transfer (IBWT) projects, which are normally backed by political pressure. These efforts occur even while efficient and effective use is not being made of the available water resources. In addition, water consumption in the urban, agricultural, and industrial sectors is not satisfactory. Although the water transfer projects have helped to reduce the intensity of water shortage in the receiving areas, a holistic and integrated approach has not been observed in planning for these projects.

Recent years have witnessed a renewed emphasis on integrated management of water resources, on the basis of which a new comprehensive water act and water management system is being developed. Increased awareness among water managers of the advantages of integrated water resource management and increased concerns about the possible adverse effects of IBWT projects have led to more careful examination and more in-depth study of new proposals.

This paper will present the current water resources situation and the trend in water consumption in Iran, and outline the new policies and strategies adopted by the water sector. It will also review and discuss information on several large interbasin water transfer projects, some implemented and some proposed for implementation. The paper will conclude with results and recommendations.

INTRODUCTION

Water is a vital natural resource for human beings and other forms of life; it is crucial to food production, serves as an input to many economic activities, and possesses spiritual and psychological values beyond its strictly utilitarian functions.

In semiarid and arid regions, limited water is available for beneficial uses. Iran is an arid country with an average annual precipitation of about 250 mm; less than one-third of the world average. Renewable water resources in Iran have been estimated to be 130,000 mm^3 per year, which may be considered the long-term average of national water assets. The increasing water demand has caused an alarming decrease in per capita renewable water available. The annual per capita water as a general index of the water resources status used to be about 7,000 m^3 in 1956 when the population was only 19 million. At present, with a population that has grown to about 65 million, the index is estimated to be about 2,000 m^3. With the increasing trend in population growth, it is predicted to sink further, to below 1,000 m^3 in the year 2025. These figures clearly show that our future generations are to face a serious water shortage during the coming decades. Pollution of water resources due to human activities makes this situation even worse.

Substantial achievements have been made during the last two decades in utilizing water resources for irrigation, domestic and industrial water supply, and hydropower generation. Due to the present increase in water demand by an increasing population, as well as the economic status and desire for higher standards of living by the public, the conflicts of water supply and water demand are gradually getting more serious so that the available water must be shared among different regions in the country for different uses. In most areas, local water resources have already been tapped, while demand remains beyond the capacity of existing water resources. Bridging the present and future gaps between demand

and supply will require tremendous efforts to develop various supply enhancements and demand management options, among which construction of new dams on the remaining unexploited sites and interbasin water transfer (IBWT) appear to water managers to be promising options. A number of IBWT schemes have accordingly been implemented or proposed in order to divert water mainly from northeastern basins into central arid regions.

Over the past few decades, IBWT projects have aroused much controversy and debate in the world and have drawn the attention of interest groups and scientists concerned with the protection of the environment and the subsequent social disruption caused by such schemes. In complex issues such as water management, where the problems and solutions differ in each and every case, the best way to understand and develop procedures for reasonable solutions is through examples. Hence, it is the aim of this paper to review some IBWT projects in Iran (though not in technical detail) and their assessment in the context of sustainable water management. In the first part of this paper, the basic concepts of IBWT and the assessment criteria are presented. These are followed by an overview of water resources and water use as well as the National Water Policy and Strategy and IBWT practices. Finally, conclusions and some recommendations are put forth.

INTERBASIN WATER TRANSFER

Interbasin water transfer is a response to a distribution of human population and related activities that differs from the spatial distribution of water resources. In arid and semiarid regions, IBWT has a long history as a means of addressing water scarcity in one region by transporting additional supplies from regions where water is relatively more abundant.

Three conditions are generally identified as necessitating water transfer systems as follows:

1. Local hydrologic conditions limit the availability of water resources;
2. Local centers of demand cannot be supplied by local resources in support of continuing urban development;
3. Local water supply conditions need improvement because of depletion and pollution of local water resources.

Interbasin water transfer is a technical (structural) tool consisting of the sharing of water resources between a receiving basin and an area of origin as the donor area. Prerequisites for successful cooperation between the receiving and the donor areas must be a shared management expected to promote sustainability, observation of common ethical values, production of benefits to both areas, and administration by suitable institutions. An unsustainable situation can be seen as the driving force for interbasin water sharing. Such an unsuitable situation happens where lack of water security and shortages reduce the quality of life, economic development, and social progress. The objective in water sharing is the betterment of living standards and social welfare. Water transfer must not be associated with unwanted and adverse effects on sustainable water management in the donor basin through environmental, social, and cultural degradation.

Large hydraulic structures such as tunnels, canals, pipes, dams, and pumping stations are commonly required in the implementation of IBWT projects. These projects also involve costly infrastructural solutions in water management. One argument often raised against infrastructural solutions is that initiatives must be directed toward reducing water wastage before any efforts are made to justify costly investments that are associated with intervening in the natural hydrologic cycle.

CRITERIA FOR EVALUATION

Reducing water shortages that are serious hindrances to the sustained development of regions creates a significant advantage of IBWT. The disadvantage, however, is that transfer may entail restrictions on future development of the donor area along with other negative effects. It follows, then, that each IBWT proposal must be thoroughly evaluated in order to establish whether or not the project is justified; in certain cases, such proposals may be totally prohibited. The following five criteria to be used in justifying or rejecting IBWT projects are proposed by W.E. Cox (Cox, 1999):

Economic Productivity Impacts

Criteria 1 and 2 involve impacts to economic productivity in both the receiving area and donor area of the proposed IBWT project:

Criterion 1: The area of delivery must face a substantial deficit in meeting present or projected future water demands after consideration is given to alternative water supply sources and all reasonable measures for reducing water demand.

Criterion 2: The future development of the area of origin must not be substantially constrained by water scarcity; however, consideration of a transfer that could constrain future development of the area of origin may be appropriate if the area of delivery compensates the area of origin for productivity losses.

Environmental Quality Impacts

Criterion 3: A comprehensive environmental impact assessment must indicate a reasonable degree of certainty that the proposed IBWT project will not substantially degrade environmental quality within the area of origin or area of delivery; however, transfer may be justified where compensation to offset environmental injury is provided.

Sociocultural Impacts

Criterion 4: A comprehensive assessment of sociocultural impacts must indicate a reasonable degree of certainty that the proposed IBWT project will not cause substantial sociocultural disruption in the area of origin or area of water delivery; however, transfer may be justified where compensation to offset potential sociocultural losses is provided.

Benefit Distribution Considerations

Criterion 5: The net benefits from transfer must be shared equitably between the area of transfer origin and the area of water delivery.

WATER RESOURCES AND USES IN IRAN

Iran occupies an area of 1,648,000 km^2, which is 1.2 percent of the earth's total area. Climatologically, Iran is located in an arid and semiarid zone with an average annual precipitation of around 250 mm, which is about 34 percent of the average annual precipitation in Asia, or about 30 percent of the average annual precipitation in the world. The volume of renewable water resources formed due to precipitation does not have an even distribution, so that 6 percent of the total land area of the country receives less than 50 mm of precipitation, 45 percent receives less than 200 mm, and only 8 percent receives between 500 to 1,000 mm of rain throughout the year. The precipitation over 1 percent of the total area is more than 1,000 mm. As seen in Figure 1, the northern and western parts of the country receive the highest amounts of precipitation while the central and eastern parts receive the lowest quantities. The country is divided into six regions making up 37 basins (Figure 2).

Table 1 shows the volume of precipitation as well as the renewable water resources (precipitation minus evapotranspiration) over the six regions of Iran. Taking into account the water entering the country from across national borders, the total renewable water amounts to around 130 billion cubic meters (bcm) (Jamab Consulting Engineers). Water consumption across Iran in 1994 is estimated to have been 87 bcm, and it is projected to increase to about 116 bcm in 2021 (Jamab Consulting Engineers).

Regarding the uneven distribution of both precipitation and population across Iran, the per capita volume of renewable water will vary from place to place. Figure 3 shows the per capita volume of renewable water over six areas in Iran according to the National Comprehensive Water Studies carried out in 1994. Based on the present population of the country, the annual average per capita volume of renewable water is estimated to be around 2,000 m^3, and it is estimated to decrease to below 1,000 m^3 by 2025. Thus, it may be predicted that within the next two decades, most parts of Iran will be facing chronic water shortage.

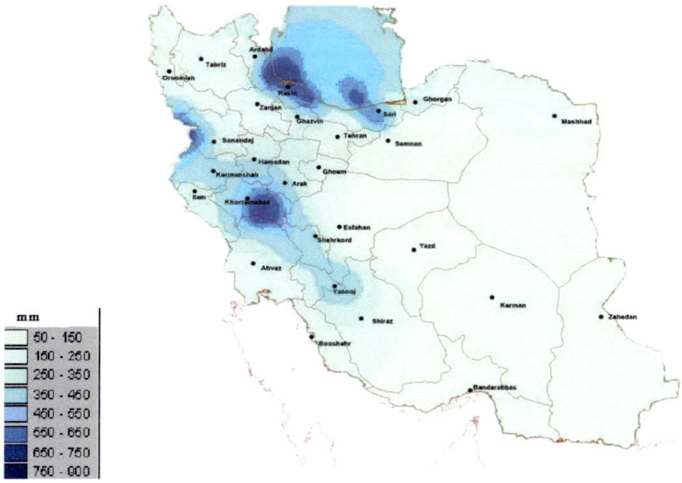

FIGURE 1 Precipitation Map of Iran (1999).

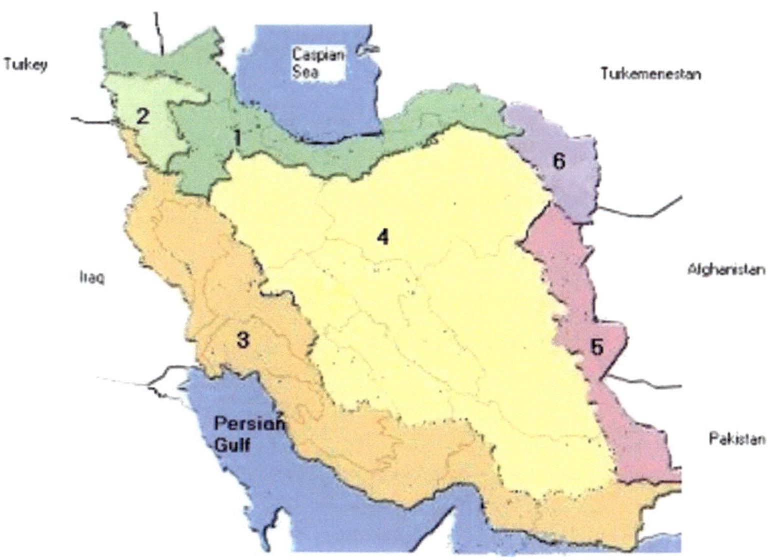

FIGURE 2 Main basins of Iran.

TABLE 1 Main Features of Regions of Iran (Jamab Consulting Engineers)

Region #	Region Name	Area (km^2)	Precipitation Volume (mm^3/year)	NPV[a] (mm^3/year)
1	Caspian Sea	173,730	84,190	24,834
2	Urumieh Lake	51,866	22,300	7,207
3	Persian Gulf	419,802	153,820	62,035
4	Central	851,126	127,510	29,584
5	Moshkil Hirmand	107,369	13,480	1,910
6	Kashaf-rood	44,107	11,860	2,430
Sum				117,000
Across border				13,000
Total				130,000

[a]NPV = Net Precipitation Volume = Precipitation Volume – Evapotranspiration.

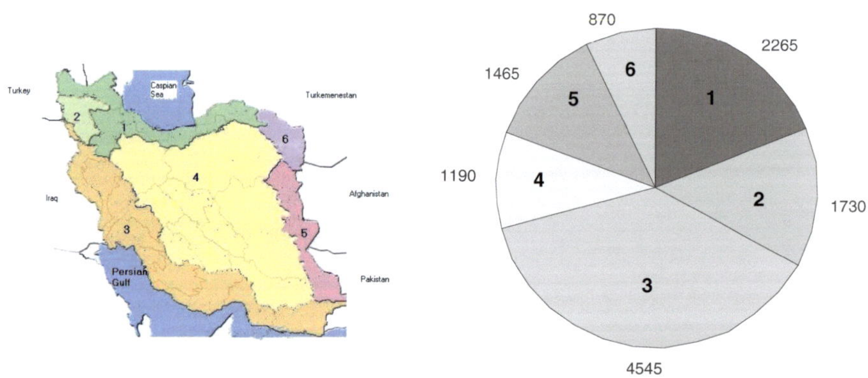

FIGURE 3 Per capita water resources (m^3/year in 1994).

Figure 3 represents the uneven distribution of water in proportion to population distribution across the country. Besides the uneven distribution of population, the distribution of agricultural, industrial, and urban activities makes uneven water distribution even more important. In fact, it can be claimed that it is not only the case that water has a higher value added in certain areas as compared to others, but that the demand for water is also far higher than its supply. There are many other underlying management causes for existing water shortages such as overexploitation of groundwater resources, pollution of surface and ground water resources, low water use efficiencies in agricultural, urban, and industrial sectors, and low water productivity in agriculture as common phenomena in the country.

THE NATIONAL WATER POLICY AND STRATEGY

In the past, water resources management in Iran did not take shape on the basis of an integrated and holistic approach. The result was that the integrity of water resources and ecosystems came under serious threat. Increased awareness of the limitations of natural resources and of the role they play in sustainable development has augmented the necessity for a shift from a nonintegrated resource development approach to a holistic, sustainable, and integrated one. Among the factors raising further motivation for this shift are the increasing gap between demand and available supply, the attendant resource degradation, and the global development and application of integrated management concepts and objectives.

Some of the macropolicies and strategies adopted in 1999 by the Council on Determining the Islamic Regime's Priorities are as follows:

- Establishment of a comprehensive management system over the whole water cycle according to the principles of sustainable land and water development over the river basins;
- Development of water resources within the framework of national plans and comprehensive river basin plans;
- Integration of water resources development, exploitation, and protection plans with other national and regional plans;
- Promotion of productivity and attention to the economic, security, and political value of water in water harvesting and extraction, supply, storage, and consumption;
- Establishing compatibility between efficiency and social justice;
- Promotion of public awareness, improving tariffs and ensuring their transparency.

The main objective in water management, for which the government takes primary responsibility, is the balancing and integrating of human relations with the process of water resources development, while taking into consideration criteria and guidelines for sustainable development.

The fundamental principles of water resources management based on the macropolicies of the water sector and the above-mentioned objectives of water management include:

- Encompassing the total water cycle in water management, in which the river basin is regarded as the main element in water management structure.
- Pursuing holistic and integrated water management, in order to establish a balance between supply and demand while taking into account the principles of sustainable development, combining structural and nonstructural measures, increasing participation of and coordinating all effective governmental and nongovernmental bodies and the various economic and social sectors.

• Promoting productivity and preventing exploitation of the economic, security, and political value of water in its harvesting, storage, supply, and consumption. Efficiency in use must be constantly improved in the various functions of water, and the motivation for conservation and efficiency must be strengthened through such measures as education and training, as well as the application of technical and economic means.

• Establishing formal water tariffs such that the motivation for economic uses of water is strengthened justly, and also such that the maintenance and operation costs and part of the investments in water projects are covered.

• Increasing development of conventional and nonconventional water resources with consideration of the drawbacks and repercussions of development of resources and its effects on third parties, also seeking to minimize both natural and man-made damage. Damage and uncertainties due to economic, social, and environmental issues must be systematically calculated and included in every development proposal.

• Coordinating land and water management within the framework of an entire watershed.

• Coordinating between various sectors in water management.

• Performing, planning, and decision-making on interbasin water transfer projects within an integrated and holistic management framework.

• Establishing and strengthening local management within the water management structure (such as national water councils, river basin councils, and local water councils).

In relation to IBWT, the fundamental principles of water resources management explicitly state that, "Planning and decision-making about interbasin transfers or transfers over large distances must be made within the framework of demand management and on the basis of criteria for minimizing costs, productivity of service, and efficiency in water uses."

According to the macropolicies and strategies in the water sector and the fundamental principles of water resources management, a comprehensive water act is in the process of development. Once this act is developed and adopted by the respective authorities, it is expected that water resource development projects, including IBWT projects, will be prepared and developed within the framework of the strategies and the fundamental principles of integrated and comprehensive water management. Nevertheless, there are many social, political, institutional, and legislative obstacles to an effective water sector reform which need realistic assessment and provisions.

INTERBASIN WATER TRANSFERS IN IRAN

Uneven water distribution and demand across the country and the shortage of water in some areas have caused water managers to consider interbasin water transfer as a solution to these water problems. Most of these projects concentrate on the transfer of water from Regions 1 and 3 to Region 4. Table 2 presents the major interbasin water transfer projects.

Figure 4 shows the interbasin water transfer activities over the three Five-Year National Development plans (1989-2003) using the index W×L, where W is the volume of water transferred and L is the transfer length or distance.

In consideration of the fact that the two significant factors affecting costs and environmental impacts in any IBWT project are the volume of water transferred (W) and the transfer length (L), Shiklomanov (1999) proposed a combination index of WxL (Km^3/year-km) to be used in the classification of IBWT projects, i.e., the product of annual transfer volume by transfer length (Shiklomanov, 1999). The classification of IBWT projects in Iran according to this index is presented in Table 3, where it is seen that most projects in Iran are either "small" or "very small."

From among the IBWT projects implemented in Iran, descriptions on the following projects are provided in this paper: water transfer from the Karoon and Dez basins to the Zayanderud basin, Intrabasin Water Transfer to the City of Zahidan, and the Northern Iran Interbasin Water Transfer Project (Mazandaran Canal), each with its own distinctions.

TABLE 2 Key Interbasin Transfer Schemes

Project	Source Basin/ Region[a]	Recipient Basin or City/Region[a]	Purpose[b]	Status[c]
Kuhrang-1	Karoon(3)	Zayanderud(4)	A, M, I	1954
Kuhrang-2	Karoon(3)	Zayanderud(4)	A, M, I	1985
Kuhrang-3	Karoon(3)	Zayanderud(4)	A, M, I	U.C.
Ch. Langan	Dez(3)	Zayanderud(4)	A, M, I	U.C.
Dez-Ghom	Dez(3)	Ghomrood(4)	M	U.C.
Soolegan	Karoon(3)	Rafsanjan(4)	A	U.I.
Halilrood	Halilrood(4)	Kerman(4)	M	U.C.
Yazd	Zayanderud(4)	Yazd(4)	M	1999
Kashan	Zayanderud(4)	Kashan(4)	M	U.C.
Talighan	Sefidrud(1)	Tehran-Ghazvin(4)	M, A	U.C.
Lar	Haraz (1)	Tehran(4)	M	1984
Zahidan	Hirmand(5)	Zahidan(5)	M	U.C.
Tabriz	Miandoab(2)	Tabriz(2)	M	1999
Moharram	Karun(3)	Persian Gulf Cities(3)	M	U.C.
Chalos	Gilan(1)	Mazandaran(1)	A	U.C.
Mazandaran	Mazandaran(1)	Golestan(1)	A	U.I.

[a]Region is shown in parentheses following each Basin or City name.
[b]A = Agricultural, M = Municipal, I = Industrial
[c]U.C. = Under Construction, U.I. = Under Investigation

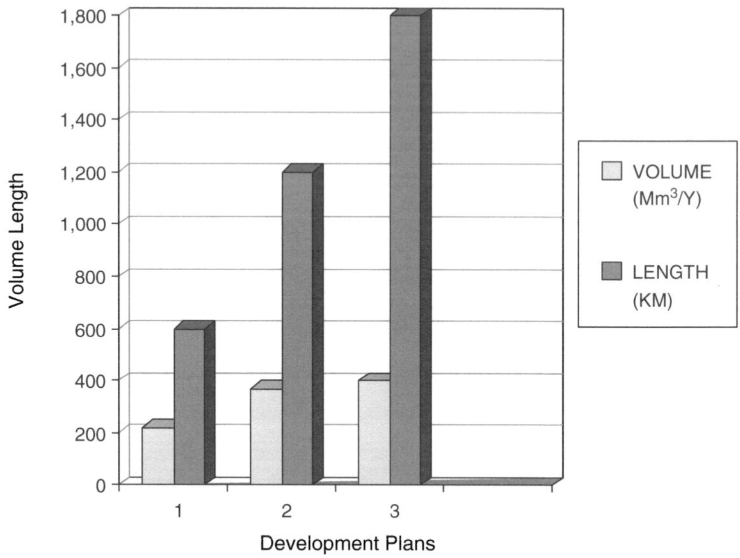

FIGURE 4 Interbasin water transfer projects over the three Five-Year Development Plans (1989-2003), measured in terms of volume of water transferred and length, or distance, of water transfer.

TABLE 3 Specification of Key IBWT Schemes

Project	Length (km)[a]	Volume (Mm^3/y)	Shiklomanov Category[b]
Kuhrang-1	2.8 (T)	300	V.S.
Kuhrang-2	2.8 (T)	160	V.S.
Kuhrang-3	23.4 (T)	260	V.S.
Ch. Langan	15 (T)	200	V.S.
Dez-Ghom	230 (P,T)	340	S.
Solegan	438(P,T)	200	S.
Halilrood	95 (P,T)	75	V.S.
Yazd	335 (P)	90	S.
Kashan	190 (P,T)	42	V.S.
Tabriz	180 (P)	150	S.
Zahidan	200 (P)	30	V.S.
Talighan	10 (T)	450	V.S.
Lar	23(T)	200	V.S.
Persian Gulf Cities	744 (P)	182	M.
Chaloos	111(C)	280	S.
Mazandaran	450(C)	950	M.

[a]Type of infrastructure is indicated in parentheses following the length: T = Tunnel, P = Pipe, C = Canal
[b]S= Small, V.S.= Very Small, M.= Medium

INTERBASIN WATER TRANSFERS FROM THE KAROON AND DEZ TO THE ZAYANDERUD BASIN

The Karoon and Dez basins are among the largest basins in Iran. The Karoon River has the largest discharge among the rivers in Iran and the Dez stands second to it. These two rivers stretch over a long distance to join the Arvandrood and, finally, the Persian Gulf in the south. There are several large dams on these rivers, and some more are either under construction or under study.

The Zayanderud River basin, with a total area of 41,542 km^2, is located in a dry and hot zone in central Iran. The average annual precipitation in the basin is estimated to be 245 mm, which varies from 50 mm in the East to over 1,700 mm in the western tributaries. The river stretches over a distance of 355 km and passes through the historic city of Isfahan. The multipurpose Zayanderud reservoir was constructed in 1970, 110 km west of Isfahan. The river supplies water for the various agricultural, industrial, and urban needs of the areas and towns located within the Zayanderud River basin as well as several towns outside the basin. Demographically, the Zayanderud River basin hosts a rather high population and is the target of many immigrants due to its location in central Iran, the presence of major industries in the region, its fertile land, as well as its scientific, tourist, cultural, and historical attractions. Several IBWT projects transferring water from adjacent rich basins (i.e., the Karoon and the Dez) have been implemented or are under construction in order to strengthen and augment the water resources in the basin (Kuhrang Tunnels I, II, and III, and Cheshmehlangan Tunnel) (Figure 5).

Central Iran is host to the provinces of Isfahan, Yazd, Kerman, Markazi, and Qom, all of which are considered to be arid areas suffering from water shortage. The IBWT projects implemented or under construction are mainly aimed at supplying water to the major cities in the region (Figure 5). Some specifications of these projects are presented in Tables 2 and 3.

MAZADARAN CANAL PROJECT

The Province of Mazandaran is one of the northern Iranian provinces located along the coast of the Caspian Sea. Land and water studies in the region indicate that the central and western parts of the province have an excessive water supply so that a considerable volume of the water remains unused and is discharged into the Caspian Sea. Considering the limited potential for agricultural development in these areas, significant increases in future water demand seem unlikely. However, in the eastern parts of the Province of Mazandaran and in the neighboring Province of Golestan, vast areas with the potential for agricultural development have remained undeveloped as of yet, due to limitations in the water resources available. Dry farming is the prevailing practice in most of these areas. A Water Resources Survey has recently been carried out by Mahab Ghods Consulting

FIGURE 5 Interbasin transfers from Persian Gulf Basins to Central Basin.

Engineers in order to determine the surplus water in the western and central parts of the Province of Mazandaran that will be transferred to Eastern Mazandaran and the Province of Golestan after environmental requirements are satisfied (Mahab Ghods Consulting Engineers); The study area stretches from the eastern banks of the Aleshrood (west of the Haraz River) to the border between Iran and Turkmenistan to the east of the Caspian Sea. In the north, the study area is bounded by the Caspian Sea, and in the south it is bordered by mountains and plains (Figure 6).

The objective of the Mazandaran Canal is to collect the surplus river water along its route to be transferred for agricultural development in the Golestan Plain. Estimates of surplus water have been based on the donor area's demand for agriculture, domestic use, industry, aquaculture development, and environmental requirements projected to the end of the planning horizon (2021). The canal in question is the continuation of another canal, currently under construction, which is intended to transfer water from the Chaloos River into Haraz Plain and to discharge the surplus water to the Mazandaran Canal (Figure 6).

According to preliminary studies, the canal length, the annual volume of water transferred, and the canal's construction costs are estimated to be 450 km, 800-1,100 Mm^3, and around 3,500 billion Rials, respectively (Mahab Ghods Consulting Engineers). The canal is planned to be an earth canal but its capacity

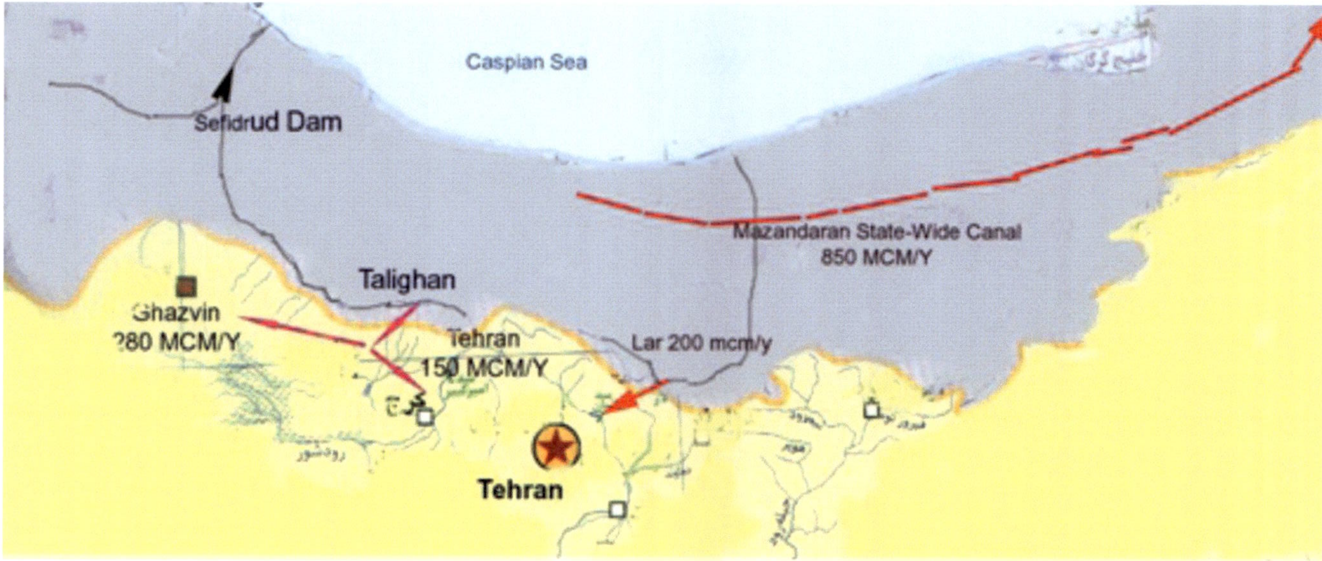

FIGURE 6 Interbasin transfers from Caspian Basins to Central Basin (Tehran).

has not yet been determined. Due to the complexities and peculiarities of the project, it was necessary to carry out more detailed studies on all hydrological, agricultural, environmental, ecological, social, and economic aspects, which are still in progress. Planning for the project is to be accomplished using multi-criteria decision-making procedures.

Some of the issues raised in relation to this project, to be explored further in the more comprehensive study, include the following:

1. Removal of trees and the disruption of rice paddies along the route.
2. Interference with fish egg-laying habits due to reduced river flow caused by water transfer into the canal.
3. Saline water intrusion into the rivers due to reduced river flow.
4. Interference with north-south traffic in the province, especially with regard to the high population density.
5. Potential hazard to local people of drowning in the canal.
6. High costs of the project.
7. High sensitivity of the project results to the assumptions and the accuracy of the study process.

THE ZAHIDAN INTRABASIN WATER TRANSFER

The city of Zahidan, the capital of Sistan and Balouchistan Province, is one of the cities facing serious problems in terms of both quality and quantity of water. The region is characterized by a desert to semi-desert climate with long, dry, and hot summers but short, dry, and cold winters, where evaporation potential is high and precipitation is low.

Zahidan has two separate water supply systems, one supplying sanitary water (saline) and the other supplying drinking water. Currently, sanitary water for the city is supplied by an old water supply system, with which water is extracted from groundwater resources. Drinking water is supplied by a small system with public valves at several points across the city. The city of Zahidan has no wastewater collection system (it is under construction), and the wastewater is discharged into pits. The groundwater resources are thus used both for public water extraction and wastewater disposal. Population growth and the corresponding increased water extractions, along with the drought over recent years, have resulted in the decline of the water table and increased water salinity. The average electrical conductivity (EC) of the groundwater within the city is currently around 4,000 microsiemens, and in some areas it has increased up to 7,000 microsiemens. This has occurred while the biological quality of the groundwater has also deteriorated due to recharge by wastewater disposal.

The sanitary supply system is around 35 years old. Long service and the excessive salinity of the flow have weakened the system. On top of this, only 65 percent of households are covered by the water supply service. The drinking water distribution system is very small and limited, and public valves deliver water to the people. The deficiencies of the water distribution system and the shortage of water in recent years have resulted in repeated interruptions and failures in supplying drinking water for long stretches of time. Therefore, most people obtain their drinking water from vendors who deliver water in tankers. The tankers take fresh water from a number of wells located to the west and southeast of Zahidan to distribute in the city. In addition to the health problems associated with this method of water delivery, overuse has recently caused the discharge from these wells to drop drastically. Due to the persistent water shortage, the per capita water consumption in the city has fallen from 160 liters per day in 1976 (both sanitary brackish water and fresh drinking water) to 110 liters in 2001.

To meet the city's water demand in the coming years, a water transmission project is under implementation that will convey water from Chahnimeh Reservoir over a route of 200 km with a pumping head of 1800 m. It will carry an annual volume of 28 mm^3 of water from the Sistan River (Figure 7).

The total cost of the transmitted water is estimated at 5,000 Rials per cubic meter (U.S. $1 equals 8,000 Rials). This will increase to 10,000 Rials once the costs of the construction of the new water supply system are added. This is based on the average price of one cubic meter of water in the city as being around 400 Rials, while the average in Iran is around 600 Rials.

As the commissioning and operation of the project draws nearer, certain issues are emerging that await solution. These include the method of distributing

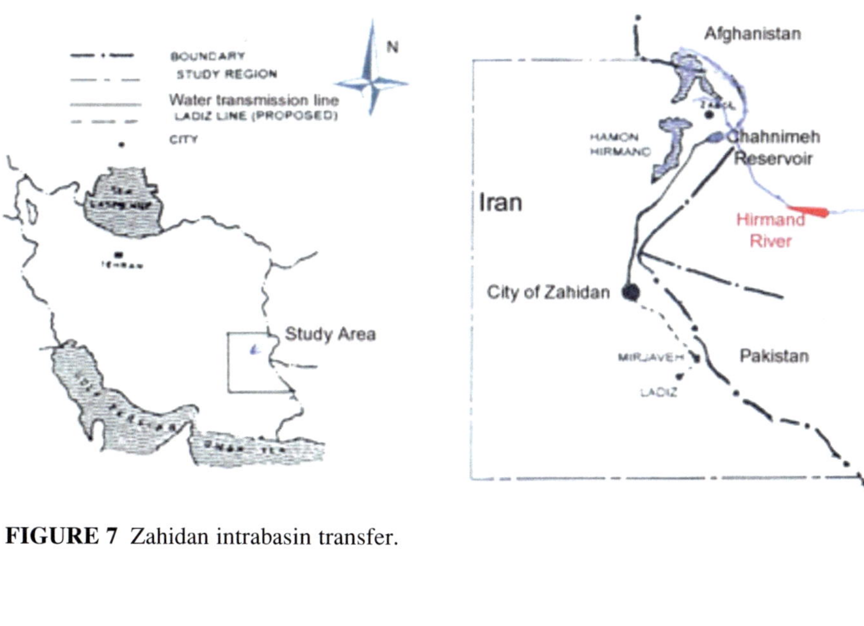

FIGURE 7 Zahidan intrabasin transfer.

water among users, and consumption patterns in the city, as well as problems arising after water is conveyed to the city (such as the expected abrupt population increase or a sudden increase in per capita consumption). It is worth mentioning that Zahidan Water & Sewage Company is in the process of designing and implementing a new water supply system to distribute the transmitted water. A per capita daily consumption of 100 liters is assumed in the design. The Company is also considering the rehabilitation and development of the two existing water supply systems. Two problems seem to arise in this system. First, the practical and possible mechanisms of controlling the per capita consumption at 100 liters/day are not clearly defined. Second, assuming a per capita consumption of 100 liters/day, the volume of water transmitted from the Chahnimeh Reservoir will not meet the urban water demands by 2011, even in the most optimistic scenarios. It is estimated that the water transmitted from the Chahnimeh Reservoir will barely suffice to meet the water demands of the city in the next 10 years. Therefore, it is essential to plan for the management of the transmitted water and the maximum use of groundwater resources. The objective in this planning must be the sustainable supply of water for the city such that the water shortage problem will be delayed to the greatest extent possible.

Planning to face these issues and problems requires comprehensive urban water management studies prior to the operation of the system under construction. The authors investigated the method of distributing the transmitted water and the existing wells, using a multicriteria decision-making method (compromise programming) (Abrishamchi et al., 2001). The general results are summarized below:

1. The results indicate that the option of bottled water deserves greater consideration in future decision-making processes. It must be emphasized that the issue definitely awaits more careful scrutiny and detailed study, especially with regards to the subsidies allocated to the water sector and its impacts on the evaluation of different alternatives.

2. In view of the sustainability criterion, future decisions must be directed toward the development and rehabilitation of the existing sanitary supply systems, construction of a drinking water supply system with public water extraction valves (at various points throughout the city but in a more systematic manner), and, finally, toward the delivery of bottled water. Indeed, the authors' final recommendation is for more detailed investigation of the above options and the development of new alternatives accordingly.

RESULTS AND DISCUSSION

Using the assessment criteria proposed by Professor Cox (Cox, 1999) for the assessment of IBWT projects, the assessment of the interbasin water transfer projects in Iran is as follows:

Criterion 1: The receiving areas have suffered from great water shortage, but the methods of reducing demand have not been properly investigated.

Criterion 2: As the Karoon and Dez basins enjoy plenty of water, the transferred volumes make up only a small portion of the available water in these basins, and the transfers do not seem to cause any water shortages or limitations on future developments in the donor area. However, since most of the upstream tributary flows located in high mountainous areas originate from snow melts or springs, the impacts of extractions from these tributaries, however small, on the reduction of the base flow in the two Karoon and Dez Rivers may need more careful study.

Some of the IBWT projects have caused shortages in irrigation water in the donor areas. Nevertheless, the government will compensate for these losses through the implementation of solution schemes. For instance, the transfer of water from the Lar Dam to Tehran draws upon the water shares of the farmers in the Haraz Plain. The construction of the Chaloos-Babol Canal with an annual transfer capacity of 280 mm^3 is expected to compensate for the water losses in the plain. The construction of the Taleghan Dam and the transfer of its water to Qazvin and Tehran will cause shortages of irrigation water in the agricultural lands downstream of Sepidrood Dam. In this case also, the government is planning for the construction of the Astoor Dam on the Ghezel Ozon River, which will contribute to a higher transfer volume from the Taleghan Dam to Tehran.

Criterion 3: No negative environmental impacts have been reported; however, a negative environmental and ecological impact is likely to occur in new projects such as the Mazandaran Canal. The possibility for such an impact is under study by the consulting engineers.

Criterion 4: No negative impacts or serious socioeconomic tensions have been reported; yet at the same time, during dry spells or dry years, objections are reported by the people in the areas of origin. The government compensates for their losses or damages through financial arrangements and compensatory schemes. In addition, water resource development schemes are also under study to supply substitute water in these areas. It also needs to be mentioned that dissatisfaction by people in donor areas is sometimes psychological and unfounded.

Criterion 5: In IBWT projects, evaluation of the benefits and the determination of fair sharing between donor and receiving areas have not been accomplished. In the absence of equitable regulations for interbasin water transfer schemes, some of the schemes have drawn heavily upon the political or power manipulation behavior of the Members of Parliament (MPs) or local water managers.

On the whole, IBWT schemes in Iran do not seem to be unacceptable options for reducing water shortages in receiving areas. However, the major defect of such schemes is that they are operated under conditions where the water consumption efficiency is still very low and the wastage of water is rather high (irrigation efficiency is around 30 percent and the urban unaccounted-for water is next to 30 percent). In order to reduce demands, water wastage, and losses, it is essential to reconsider the application of demand management and water reuse. This is mainly because IBWT schemes are only mid-term remedies for alleviating water shortage and serve only to diminish any motivations on the part of water managers or the public to implement the more sustainable solutions of demand management. This claim is borne out by the fact that it is only during droughts that managers and the public give serious attention to reducing water consumption. Demand management is not only economically cheaper but also contributes more drastically to the sustainability of water resources and environmental pollution control.

The implementation of IBWT schemes, particularly for urban water supply, where IBWT helps to ease water crises, should not lead to the neglect of limitations of water resources in urban planning. IBWT options in alleviating urban water shortages must be employed only within holistic and integrated planning and management. The mechanisms of demand management, the different methods of urban water distribution and supply (such as the use of dual networks or bottled water), utilization of small local water resources, conjunctive use of surface and ground water and urban wastewater management, reclamation, and reuse all must be considered in the integrated management approach. Urban groundwater also must be protected both quantitatively and qualitatively for emergency conditions during which the water transfer system may temporarily fail.

RECOMMENDATIONS

The following planning strategies and principles may be recommended for consideration by managers in any water transfer scheme:

1. Only objective arguments for a water transfer scheme should be presented. Explanations must be provided for the gaps between demand and supply in the receiving region, and alternative sources for fresh water supply (especially small local sources) need to be introduced.

2. Prior to the development of any IBWT scheme, the need for water transfer should be minimized. The following measures may be used in the process of minimizing the need: (i) alternative conventional and unconventional sources of water must be considered; (ii) innovative demand management measures for water conservation and saving should be effected; (iii) by improving the operation of existing water supply systems, water supply reliability should be enhanced; and (iv) alternative schemes of water distribution and delivery in urban areas (e.g., dual networks, bottled water, and other options) must be considered.

3. The water being transferred from a basin should be surplus water after meeting all the present and future reasonable foreseeable needs of the donor basin; in other words, the present and future requirements of the donor basin must be fully met or secured.

4. Prior to the implementation of IBWT proposals, thorough assessments of social and environmental impacts must be made, as well as provision for the continuous monitoring of likely impacts during construction and operation of the projects.

5. Care must be taken to minimize adverse impacts due to water transfer both in the donor and receiving areas, as well as changes in the hydrological regimes, ecological regimes, environmental pollution, and aesthetic and human interests.

6. The decision makers and the general public need to be provided with unbiased information through professional analysis and modeling tools employed to support the IBWT multicriteria decision-making process.

ACKNOWLEDGMENTS

The authors wish to thank the U.S. and I.R. Iran Academies of Science and Sharif University of Technology for supporting this work. They would also like to thank the Environment and Water Research Center (SUT), Mr. M. Zarghaami, Mr. E. Roostazadeh, and Mrs. M. Kheirabadi for their assistance; and Mr. A. Salavitabar and Mr. A. Naghibi of Mahab Ghods Consulting Engineers, and Jamab Consulting Engineers for providing valuable information and data for this paper.

REFERENCES

Abrishamchi, A., A. Ibrahimian, and M. Tajrishy. 2001. Multi-criteria decision making in urban water management, Case study: City of Zahidan. Proceedings, 2nd Asian Conference on Water and Wastewater Management, Tehran, Iran.

Cox, W.E. 1999. Determining when interbasin water transfer is justified: Criteria for evaluation. Proceedings, International Workshop on Interbasin Water Transfer, UNESCO, Paris, April 1999. Pp. 173-178.

Jamab Consulting Engineers National Water Master Plan, Water Resources Organization of Iran, 1378 (in Farsi).

Mahab Ghods Consulting Engineers. Mazandaran Canal, Technical Report. Iran Water and Power Resources Development Company, 1380 (in Farsi).

Shiklomanov, I.A. 1999. Water transfer as one of the most important ways to eliminate water resources deficits and to solve water management problems. Proceedings, International Workshop on Interbasin Water Transfer, UNESCO, Paris, April 1999. Pp. 203-210.

APPENDIXES

Appendix A

Workshop Agenda

WATER CONSERVATION, REUSE AND RECYCLING

Tunisia, December 9-15, 2002

Monday, December 9—Tunis

Morning site visits

Tunisia International Centre for Environment Technology.(CITET). Presentations of the water resources mobilisation and exploitation strategy in Tunisia by the General Department of dams and great hydraulic works (DGBGTH), and presentation of the Tunisia International Centre for Environment Technology activities.

Northern Water Open Channel and Pipelines Exploitation Company (SECADENORD) at Bejaoua. Presentation of northern water management strategy.

National Society for Water Exploitation and Distribution (SONEDE) at Ghedir-el-Golla. Presentation of the drinking water treatment station.

Afternoon site visits

National Improvement Office (ONAS) at Soukra (Chotrana). Presentation of the used water treatment station.

Tuesday, December 10—Golden Tulip Hotel

Morning session

Introduction to Opening Session

Presentation: General Issues Related to Agricultural, Urban, and Industrial Water Conservation and Reuse in the U.S.

Presentation: Overview of Water Management, Conservation, and Reuse in Iran

Roundtable Discussion of General Issues Related to Agricultural, Urban, and Industrial Water Conservation and Reuse

Afternoon session

Presentation: Treatment Technologies I: Large Scale Systems

Presentation: Treatment Technologies II: Small and Decentralized Technologies

Presentation: Desalting by Solar Energy

Round Table Discussion of Afternoon Presentation

Wednesday, December 11

Morning session

Presentation: Overview of Cultural Practices for Optimal Irrigation in the Semi-Arid U.S.

Presentation: Status of Water Conservation and Reuse in Agriculture in Iran

Managing Drought in the Agricultural Sector

- Case Studies from Iran
- Case studies from the western U.S.

Afternoon session

Round Table Discussion to Identify Promising Technologies, Practices, and Priorities for Optimal Irrigation

Presentation: the Economics of Agricultural Water Use and the Role of Prices

Presentation: Problems of Water Allocation and Pricing in Iran

Round Table Discussion of Issues and Problems with Allocation and Pricing

Discussion of Opportunities for Collaboration

Thursday, December 12

Morning session

Presentation: Case Studies from the U.S. on Municipal Conservation and Reuse

Identification and Discussion of Lessons and Principles from U.S. Experience with Municipal Conservation and reuse

Presentation: Identifying Microbial and Chemical Contaminants for Regulatory Purposes: Lessons from the U.S.

Afternoon session

Round Table Discussion to Identify Important Principles and Priorities: The Lessons

Presentation: Interbasin Water Transfers in the Western U.S.: Issues and Lessons

Presentation: Interbasin Water Transfers in Iran

Round Table Discussion of Issues and Problems of Interbasin Transfers and Identification of Additional Opportunities for Collaboration

Friday, December 13—Hasdrubal Hotel, Djerba

Morning session

Presentation of three conferences at the hotel on:

- The strategy of water resources integrated management—*General Department of Dams and Great Hydraulic Works.*
- The national cleaning up strategy—*National Improvement Office (ONAS).*
- The national strategy of using treated water for agricultural irrigation—*General Department of Rural Engineering and Water Exploitation.*

Travel to Djerba. Presentation of the water desalination station in Djerba. *SONEDE.*

Afternoon session

Travel to the Matmata Hotel. Discussion on conferences and visits—*CRDA of GABES.*

Saturday, December 14—Palm Beach Hotel, Tozeur

Morning session

Presentation of three conferences at the hotel on:

- National strategy of soil and water conservation—*General Direction of Development and Agricultural Land Conservation.*
- National strategy of water saving and water pricing—*General Department of Rural Engineering and Water Exploitation*
- Geothermal agriculture—*Regional Representative for Agricultural Development at Kebill.*

Site visits—*CRDA of Tozeur.*

- Tamaghza oasis.
- el-Oudi dam.
- Midas oasis.

Afternoon session

Travel to Tamaghza Palace hotel. Discussion about the conferences and visits.

Visiting the geothermal water irrigated parcel at el-Hamma.

Sunday, December 15

Morning session

Site visit to the treated water irrigated perimeter region at Sfax—*CRDA of Sfax*

Afternoon session

Return to Tunis. Closing ceremony of the workshop and exchanging experiences between the American, the Iranian, and the Tunisian delegations.

Appendix B

Workshop Participants List

PARTICIPANTS LIST

Henry Vaux, *Chair,* Associate Vice President, University of California
Mehdi N. Bahadori, Vice-President for Research, Iranian Academy of Sciences
Abbas Afshar, Head of the Department of Civil Engineering, Iran University of Science and Technology
Amin Ali Zadeh, Professor of Irrigation, College of Agriculture, Fedwosi University of Mashhad
James Crook, Principal Water Reuse Technologist, CH2MHILL
David Getches, Rafael Moses Professor of Natural Resource Law, University of Colorado School of Law
Abbas Keshavarz, Agricultural Irrigation Scientist, Agricultural Research, Education, and Extension Organization (AREEO)
Stephen Lacy, MWH Americas, Inc.
John Letey, Professor, Department of Environmental Sciences, University of California
Seyed Mousavi, Professor of Water Resources, College of Agriculture, Isfahan University of Technology
Kara Nelson, Professor, Department of Civil and Environmental Engineering, University of California
Rebecca Parkin, Professor, School of Public Health and Health Services, The George Washington University Medical Center
David Sunding, Professor, Department of Agriculture and Natural Resource Economics, University of California
Mohammad Hossein Tajrishi, Assistant Professor, Civil Engineering Department, Sharif University of Technology